SUPERCONDUCTING ACCELERATOR MAGNETS

SUPERCONDUCTING ACCELERATOR MAGNETS

K.-H. Mess
DESY, Germany

P. Schmüser
Universität Hamburg and DESY, Germany

S. Wolff
DESY, Germany

World Scientific
Singapore • New Jersey • London • Hong Kong

Published by

World Scientific Publishing Co. Pte. Ltd.

P O Box 128, Farrer Road, Singapore 912805

USA office: Suite 1B, 1060 Main Street, River Edge, NJ 07661

UK office: 57 Shelton Street, Covent Garden, London WC2H 9HE

British Library Cataloguing-in-Publication Data
A catalogue record for this book is available from the British Library.

SUPERCONDUCTING ACCELERATOR MAGNETS

ISBN 981-02-2790-6

This book is printed on acid-free paper.

Printed in Singapore by Uto-Print

Preface

It is the purpose of this book to collect the experience gained with the design, construction and operation of the superconducting magnets for large hadron accelerators, and to outline the physical principles of superconductivity and its application in high-field dipole and quadrupole magnets. Most of the material stems from the extensive research and development work at the large high energy physics laboratories and of course also from the authors' own work at the proton-electron colliding beam facility HERA in Hamburg. After the success of the magnets for the Tevatron proton accelerator at the Fermi National Accelerator Laboratory near Chicago, a number of new projects have been initiated and the community of physicists and engineers working in this field has considerably expanded. Many of those people, the authors included, have profited a great deal from the excellent book *Superconducting Magnets* by Martin N. Wilson. Also the book *Superconducting Magnet Systems* by H. Brechna is an extremely useful reference. It is by no means our intention to replace these monographs, on the contrary we try to be complementary and focus our attention on the aspects that are specific to accelerator magnets. There is of course some unavoidable overlap, for instance in field computation, the treatment of persistent magnetization currents or quench propagation but also here we concentrate on recent experimental results and their interpretation.

Field quality is one of the key issues for the magnets of a hadron collider. The impact of mechanical tolerances, persistent currents and eddy currents on field quality is discussed at length and numerous experimental data are presented. Another key issue is the quench performance of the magnets. Unlike in many other applications of superconducting magnets, for example in magnetic resonance imaging, quenches in accelerator magnets cannot be avoided, because beam losses happen from time to time. The quench detection and magnet protection system is hence of vital importance and needs considerable attention.

Within the scope of the book it is impossible to fully cover the enormous amount of research and development work done at the Lawrence Berkeley Laboratory (LBL), Brookhaven National Laboratory (BNL), Commissariat a l'Energie Atomique (CEA) in Saclay, the European Laboratory for Particle Physics (CERN), Deutsches Elektronen-Synchrotron (DESY), Fermi National Accelerator Laboratory (FNAL or Fermilab), the Japanese Centre for High Energy Physics KEK, Rutherford Laboratory, the former Superconducting Super Collider Laboratory (SSCL) and other laboratories. We apologize for having to omit important developments. Data from magnetic and other measurements are presented to illustrate the properties and performance of practical superconducting dipoles and quadrupoles as well as the implication of superconductor-related properties for the particle beams. Again a selection had to be made, we hope it was fair.

When the HERA project was started we received generous help and advice from scientists and engineers at Fermilab, Brookhaven, CERN and Saclay: A.V. Tollestrup, R. Lundy, J. Carson, J.G. Cottingham, R.B. Palmer, W.B. Sampson, D. Leroy, J. Perot, G. Tool and many others. During the development and production phase of the HERA magnets it has been a very rewarding experience to work under the skillful leadership of B.H. Wiik whose determination and constant encouragement were vital for the successful outcome. The technical expertise of D. Degèle, G. Horlitz, H. Kaiser and R. Meinke has been extremely valuable. The close and fruitful collaboration with the physicists, engineers and technicians in the DESY magnet design and magnet measurement groups and the cryogenic group is greatly appreciated.

We have profited a great deal from numerous discussions with our colleagues and friends at other laboratories and thank for their advice: M. Garber, A.K. Ghosh, M. A. Green, D. Hagedorn, M. N. Wilson. Special thanks go to A. Devred, A.F. Greene, R. Perin and P. Wanderer for providing us with a wealth of information on the SSC, RHIC and LHC magnets. A. Devred read several chapters in advance and made very valuable suggestions.

We thank Mr. D. Kahnert, Ms. I. Nickel and Ms. V. Werschner for their help in the layout of the manuscript and Ms. E. Dinges and Ms. M. Glänzel for preparing the drawings.

Hamburg, March 1996   K.-H. Mess, P. Schmüser, S. Wolff

Contents

Chapter 1

Introduction

1.1 Field properties of superconducting magnets

The vanishing electrical resistance of superconducting coils as well as their ability to provide magnetic fields far beyond those of saturated iron is the main motivation to use superconductor technology in all new large proton, antiproton and heavy ion accelerators. The first machine of this kind, the Tevatron at the Fermi National Accelerator Laboratory near Chicago, USA, has been operating as a synchrotron and a proton-antiproton collider for many years, featuring centre-of-mass energies of 1800 GeV with excellent luminosity. The successful dipole and quadrupole magnets developed at FNAL have strongly influenced most later designs of superconducting accelerator magnets. The first dipole prototypes built at DESY, as well as similar magnets constructed at Saclay for the Russian UNK project, were basically copies of the Fermilab magnet. Essential features like the methods to wind, cure and clamp the coils have been retained in more recent designs such as in the magnets for the proton-electron collider HERA at DESY, the Superconducting Super Collider SSC in the USA[1], the large Hadron Collider LHC at CERN and the Relativistic Heavy Ion Collider RHIC at Brookhaven. Superconductivity does not only open the way to much higher particle energies but at the same time leads to a substantial reduction of operating costs. In the normal-conducting Super Proton Synchrotron SPS at CERN a power of 52 MW is needed to excite the machine to an energy of 315 GeV while at HERA a cryogenic plant with 6 MW electrical power consumption is sufficient to provide the cooling of the superconducting magnets with a stored proton beam of 820 GeV. Hadron energies in the TeV regime are practically inaccessible with standard technology.

A high field quality is needed if one wants to store an intense particle beam for many hours. The relative deviation from the ideal dipole or quadrupole field should not exceed a few parts in 10^4 to guarantee a reasonable beam intensity for storage times of 10 - 20 hours. This poses no particular problem with normal magnets whose field distribution is determined by accurately shaped iron yokes. In a superconducting

---

[1]The SSC project was canceled by decision of the United States Congress in October 1993.

coil, however, the field pattern is governed by the arrangement of the current conductors and a precise coil geometry is of utmost importance. The positional accuracy required for the conductors in the coil is in the 20 μm range. This precision must be maintained in spite of the huge Lorentz forces acting on the current conductors: the two halves of a dipole coil repel each other with a typical force of 10^6 N (100 tons) per metre length at a field of 5 T. The coils are confined by strong clamps, often called *collars*, which take up the Lorentz forces and define the exact geometry.

Superconducting magnets have a number of properties which are not found in normal magnets and require careful attention. Some of these will be addressed in the next sections.

1.2   Quenches, degradation, training

A quench is the transition from the superconducting to the normal state. Such a transition will invariably occur if any of the three parameters temperature, magnetic field or current density exceeds a critical value. The origin may be a conductor motion under the influence of Lorentz forces resulting in a heating of the cable by frictional energy. At high currents a motion of a few μm may be sufficient since only a tiny energy deposition, in the order of 1 mJ/g, is needed to heat the conductor beyond the critical temperature. The reason for this extreme sensitivity is the very low heat capacity of metals at low temperature ($C \propto T^3$ in the limit $T \to 0$). Liquid helium is the only substance with an appreciable heat capacity in the 2–4 K range. A good thermal contact between the superconducting cable and the helium coolant is therefore of utmost importance for the stability of the coil.

If a quench happens in a large dipole or quadrupole the current in the coil must be reduced to zero in a short time interval (typically in less than a second at 5 T) to avoid overheating and possible destruction of the normal conducting part of the coil. The protection of a single magnet is straightforward: when a quench is detected the power supply is switched off and the stored magnetic energy, which may amount to several Mega-Joules, is dissipated in a dump resistor. For a chain of magnets connected in series, however, the large inductance does not allow the current to be reduced to zero in less than a second because then dangerously high induced voltages of many tens of kV would arise. A possible solution is to bypass each magnet in the chain with a diode; if a magnet quenches, the current in the chain is decreased slowly but it is guided around the quenched coil by means of the diode. A reliable quench detection and magnet protection system is one of the most important safety features of a superconducting accelerator. It is equally important to construct the magnets in such a way that they have a high inherent stability against quenches.

Many superconducting magnets have shown a phenomenon called *degradation*: the magnet could not be excited up to the critical current of the conductor but quenched at significantly lower values. The reason may be insufficient clamping of the windings or insufficient cooling. If the windings can move slightly under the influence of Lorentz forces the magnet may exhibit *training*: the first quench occurs when a certain part of a winding starts to move; if this part is, after the motion, in a stable position the magnet can be excited to a higher current in the next attempt. The second quench will then be caused by the motion of another part of the windings. It is quite common that magnets can be 'trained' this way and finally reach the critical current of the conductor after a certain number of training steps. The minimum requirement for a magnet is then to exhibit no further training when it is cooled down and excited for a second time. Fortunately, the large accelerator dipole and quadrupole magnets can be built so well that they show little if any training and can be excited to the critical current of the superconducting cable almost in the first step. The essential criteria for such a good performance are a sufficiently high pre-stress in the coil preventing conductor motion and an optimum cooling by making the coils permeable to helium.

1.3 Persistent magnetization currents

The advantage of superconducting coils turns into a drawback at low fields. Any field variation induces bipolar magnetization currents in the superconductor which – in contrast to eddy currents in conventional electromagnets – do not decay exponentially but are persistent. They result in field distortions which may become intolerably large at low excitation. A well-known example is the sextupole component measured in all superconducting dipoles. In a machine with very low injection energy like HERA even higher multipoles play a role and require a compensation by correction coils. The persistent currents are not exactly constant but exhibit a slow, nearly logarithmic time dependence. The current in the correction coils has to be adjusted to compensate the drift. In addition to the superconductor magnetization currents there are eddy currents in the cable which may have time constants of hours and lead to field distortions and heat generation during a ramp of the magnetic field.

1.4 Iron yoke

Superconducting accelerator magnets are generally equipped with an iron yoke which, however, differs considerably from the yoke of a normal electromagnet. The yoke is a hollow cylinder mounted concentrically around the coil. It serves three purposes:

(1) The central field is increased by 10 to 40%, depending on the proximity between coil and yoke.

(2) The yoke screens the fringe field outside the magnet.

(3) The stored magnetic energy is reduced which is an advantage in case of a quench.

There has been a long debate on the relative virtues of 'warm' and 'cold' iron yokes, i.e. whether the yoke should be outside or inside the liquid helium cryostat. The presently favoured solution is a yoke inside the cryostat which surrounds the coil clamped with non-magnetic collars. Magnets of this type have been first constructed for the HERA proton storage ring. The LHC dipoles and quadrupoles are twin-aperture magnets: two coils of opposite polarity are installed in a common iron yoke. In the RHIC magnets the iron yoke itself is used to compress the coils.

1.5   Cryogenic system and materials

The magnet cryostats, transfer lines and other cryogenic equipment have to be carefully designed and built to minimize heat conduction and radiation from the room temperature environment to the liquid helium system. Even with the most advanced liquid helium plant, 1 W of refrigeration power at 4.2 K requires about 280 W of electrical power; so any improvement in the thermal insulation leads to a sizable reduction in operating costs. Fermilab has developed a low-loss cryostat for the SSC magnets with a sophisticated support system of the cold part and a radiation shield at 20 K in addition to the usual shield at liquid nitrogen temperature. Variants of this type of cryostat are used in all recent magnet designs (SSC, RHIC, LHC).

The most commonly used superconductor is niobium-titanium alloy. NbTi magnets are limited to about 6.5 T at a helium temperature of 4.4 K (pressurized normal helium) and 8.5–9 T at 1.8–2 K. The LHC design is based on magnets with NbTi conductor cooled by superfluid helium of 1.9 K. Superfluid helium features much lower viscosity and much higher heat conductivity than normal liquid helium. A drawback of the low operating temperature is that the enthalpy of the cable is an order of magnitude smaller at 1.9 K than at 4.2 K which requires greater care in limiting conductor motion during excitation of the magnets.

In principle niobium-tin (Nb_3Sn) would be a good conductor for high-field magnets but the material is very brittle and requires great effort in the coil production. In spite of these difficulties a 1-m-long prototype dipole built for LHC has recently been excited to 11 T without training quenches[2].

The materials used in a superconducting accelerator magnet have to be carefully selected to be suitable for a temperature of 2–4 K and for radiation doses of more than 10^6 Gy (10^8 rad) while also being non-magnetic. Many common structural or insulating materials are excluded by these requirements. Only a few types of stainless steel show no embrittlement and keep their low permeability when they are cooled down. One of the best common insulators, Teflon, cannot be used because it deteriorates already at radiation doses of about 100 Gy. Useful insulating materials are polyimides like Kapton[3], glass-fibre epoxy and mica.

---

[2]A. den Ouden et al., *The Nb₃Sn dipole project at the University of Twente*, internal report 1995.
[3]Kapton and Teflon are registered trade marks of Du Pont de Nemours.

1.6   Correction magnets

Any large accelerator — normal or superconducting — needs correction dipoles to adjust the orbit, and sextupole lenses to compensate the chromatic aberration of the quadrupole magnets, i.e. the momentum dependence of their focal strength. Often also a few octupoles and rotated quadrupoles are needed. In a superconducting accelerator, additional correction requirements arise. In HERA, for instance, a long string of dipole and quadrupole magnets is electrically connected in series because the number of current leads which feed the main magnet current (5 kA) into the cryogenic system has to be minimized to save on refrigeration power[4]. Hence the focal length of the main quadrupoles is fixed and an adjustment of the beam optics requires quadrupole correction coils. The sextupoles needed for the chromaticity correction serve the second purpose of compensating the persistent-current sextupoles of the main dipoles. In some cases further correction elements are needed. In the HERA proton ring, the injection energy of 40 GeV is only 5% of the nominal energy and the decapole (10-pole) fields in the dipoles and dodecapole (12-pole) fields in the quadrupoles have to be compensated by special correction coils to provide sufficient field homogeneity at low energies.

This book is organized as follows. A brief introduction into the physics of superconductivity is presented in Chap. 2, followed by a description of practical superconductors in Chap. 3. Magnetic field calculations are outlined in Chap. 4 including the effects of iron yoke and end fields. Mechanical imperfections and magnetic forces are treated in Chap. 5. Chapter 6 is devoted to one of the most important topics of superconducting accelerator magnets, persistent magnetization currents and their influence on the field pattern. Model calculations and experimental data are presented in detail and the time dependence of these effects is studied. The impact of eddy currents in the copper of the cable and other conducting materials is analyzed in Chap. 7. The wide field of superconductor stability, quench origins and propagation and magnet protection is addressed in Chap. 8. In Chap. 9 we first introduce a few important concepts of accelerator physics which are needed to appreciate the demanding requirements on field quality in hadron storage rings and illustrate this with practical examples from the HERA collider. Then various types of superconducting correction coils and some of their properties are described. Chapter 10, finally, is devoted to a short outline of practical construction and fabrication methods of accelerator magnets.

Some more specialized topics have been moved into the Appendix. Magnetic field measurement techniques are addressed in A and B; field distortions from persistent currents in beam pipe coils are computed in C; approximate relations for the critical parameters of NbTi are given in D; the properties of liquid and gaseous helium are described in E; Kapton properties at room and cryogenic temperature are listed in

---

[4]The current leads are cooled by helium gas evaporated from the liquid.

F; stainless steels and aluminium-alloy as collar materials and soft iron for the yoke are discussed in G.

General reading

H. Brechna, *Superconducting Magnet Systems*, Springer, Berlin 1973

H. Desportes, *Three decades of superconducting magnet development*, Cryogenics **34** ICEC Suppl. (1994) 47

A. Greene, P.C. Dent and C. Hallquist, *The magnet system of the Relativistiv Heavy Ion Collider*, Superconductor Industry Vol. 8, No. 4, p. 16, Rodman Publ., New Jersey (1995)

R. Perin, *State of the art in high-field superconducting magnets for particle accelerators*, Particle Accelerators **28** (1990) 147

R. Perin, *Review of R & D towards high-field accelerator magnets*, Proc. Europ. Part. Accel. Conf. Berlin 1992, Edition Frontières (1989) 289

R. Perin for the LHC magnet team, *Status of LHC programme and magnet development*, IEEE Trans. **ASC-5** (1995) 189

P. Schmüser, *Superconducting magnets for particle accelerators*, Rep. Prog. Phys. **54** (1991) 683 and in: M. Month and M. Dienes (Eds.), *The Physics of Particle Accelerators*, American Inst. of Physics (AIP) Conf. Proc. 249, 1992, p. 1099

A.V. Tollestrup, *Progress in superconducting magnet technology for accelerator/storage rings*, IEEE Trans. **NS-28** (1981) 3198

M.N. Wilson, *Superconducting Magnets*, Clarendon Press, Oxford 1983

P. Wanderer, *Status of superconducting magnet development (SSC, RHIC, LHC)*, Proc. Part. Acc. Conf. Washington 1993

B.H. Wiik, *Electron-proton colliders*, in: H.C. Wolfe (Ed.), *The State of Particle Accelerators and High Energy Physics*, AIP Conf. Proc. 92, 1982, p. 101

B.H. Wiik, *HERA: machine and experiment*, Proc. XXIV Int. Conf. on High Energy Physics, Munich 1988, Springer 1988, p. 404

S. Wolff, *Superconducting accelerator magnet design* in: M. Month and M. Dienes (Eds.), *The Physics of Particle Accelerators*, AIP Conf. Proc. 249, 1992, p. 1159

Chapter 2

Basics of Superconductivity

The unusual features of superconducting magnets mentioned in the Introduction are closely linked to the physical properties of the superconductor itself. For this reason a basic understanding of superconductivity is indispensable for the design, construction and operation of superconducting magnets. In this chapter we want to give a brief survey of the physics of the superconducting state[1] with special emphasis on the effects that are relevant for magnets. Only the traditional 'low-temperature' superconductors are treated since up to date the use of 'high-temperature' ceramic superconductors in electromagnets is rather limited. For more comprehensive presentations we refer to the excellent text books by W. Buckel (1990) and by D.R. Tilley and J. Tilley (1990).

2.1 Overview

Superconductivity was discovered in 1911 by the Dutch physicist H. Kamerlingh Onnes, only three years after he had succeeded in liquefying helium. During his investigations on the conductivity of metals at low temperature he found that the resistance of mercury dropped to an unmeasurably small value just at the boiling temperature of liquid helium. Kamerlingh Onnes called this totally unexpected phenomenon 'superconductivity' and this name has been retained since. The temperature at which the transition took place was called the *critical temperature T_c*. Superconductivity is observed in a large variety of materials but, remarkably, not in some of the best normal conductors like copper, silver and gold, except at very high pressures. This is illustrated in Fig. 2.1 where the resistivity of copper, tin and the 'high-temperature' superconductor $YBa_2Cu_3O_7$ is sketched as a function of temperature. Table 2.1 lists some important superconductors together with their critical temperatures at vanishing magnetic field.

A conventional resistance measurement is far too insensitive to establish infinite conductivity. A much better method consists of inducing a current in a ring and

---

[1]Adapted from P. Schmüser, *Superconductivity*, Lectures at the CERN-DESY School on 'Superconductivity at Particle Accelerators', Hamburg 1995, to be published.

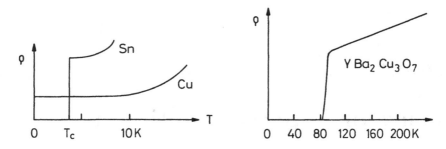

Figure 2.1: The low-temperature resistivity of copper, tin and YBa$_2$Cu$_3$O$_7$.

Table 2.1: Critical temperature of selected superconducting materials for vanishing magnetic field.

material	Al	Hg	Sn	Pb	Nb	Ti	NbTi	Nb$_3$Sn
T_c [K]	1.14	4.15	3.72	7.9	9.2	0.4	9.4	18

determining the decay rate of the produced magnetic field. The induced current should decay exponentially

$$I(t) = I(0)\exp(-t/\tau)$$

with the time constant given by the ratio of inductance and resistance, $\tau = L/R$, which for a normal metal ring is in the order of 100 μs. In superconducting coils, however, time constants of up to 10^5 years have been observed (File and Mills 1963). So the resistance must be at least 14 orders of magnitude below that of copper and is indeed indistinguishable from zero. An important practical application of this method is the operation of solenoid coils for magnetic resonance imaging in the short-circuit mode which exhibit an extremely slow decay of the field of typically $3 \cdot 10^{-9}$ per hour (M.N. Wilson, private communication).

There is an intimate relation between superconductivity and magnetic fields. W. Meissner and R. Ochsenfeld discovered in 1933 that a superconducting element like lead completely expelled a weak magnetic field from its interior when cooled below T_c while in stronger fields superconductivity broke down and the material went to the normal state. The spontaneous exclusion of magnetic fields upon crossing T_c could not be explained in terms of the Maxwell equations of classical electromagnetism and indeed turned out to be a quantum-theoretical phenomenon. Two years later, H. and F. London proposed an equation which offered a phenomenological explanation of the field exclusion but the justification of the London equation remained obscure until the advent of the microscopic theory of superconductivity by Bardeen, Cooper and Schrieffer (1957). The BCS theory revolutionized our understanding of this fascinating phenomenon. It is based on the assumption that the supercurrent is not carried by single electrons but rather by pairs of electrons of opposite momenta and spins, the so-called *Cooper pairs*. All pairs occupy a single quantum state, the BCS ground state, whose energy is separated from the single-electron states by an energy gap $2\Delta(T)$. The critical temperature is related to the energy gap at $T = 0$ by

$$1.76\, k_B T_c = \Delta(0) \ . \tag{2.1}$$

Here $k_B = 1.38 \cdot 10^{-23}$ J/K is the Boltzmann constant. The magnetic flux through a superconducting ring is found to be quantized, the smallest unit of flux being the elementary flux quantum

$$\Phi_0 = \frac{h}{2e} = 2.07 \cdot 10^{-15}\, \text{Vs} \tag{2.2}$$

where $h = 6.626 \cdot 10^{-34}$ Js is Planck's constant and $e = 1.602 \cdot 10^{-19}$ C the fundamental unit of charge. These and many other predictions of the BCS theory, like the temperature dependence of the energy gap and the existence of quantum interference phenomena, have been confirmed by experiment and often even found practical application. Unfortunately this fundamental theory is quite difficult to understand and we shall not pursue it any further. Even the most important property of a superconductor, namely its ability to carry a resistance-free current, is hard to explain in simple terms. It is related to the existence of the energy gap and a deeper understanding is only possible in the framework of quantum theory.

   A discovery of enormous practical consequences was the finding that there exist two types of superconductors with rather different response to magnetic fields. The elements lead, mercury, tin, aluminium and others are called 'type I' superconductors. They do not admit a magnetic field in the bulk material and are in the superconducting state provided the applied field stays below a *critical field* B_c which is usually less than 0.1 Tesla. All superconducting alloys like lead-indium, niobium-titanium, niobium-tin and also the element niobium belong to the large class of 'type II' superconductors . They are characterized by two critical fields, B_{c1} and B_{c2}. Below B_{c1} these substances are in the *Meissner phase* with complete field expulsion while in the range $B_{c1} < B < B_{c2}$ they enter the *mixed phase* in which magnetic field pierces the bulk material in the form of flux tubes. These materials remain superconductive up to much higher fields (10 Tesla or more) and would at first sight appear as ideally suited for magnet coils. There is, however, a disturbing phenomenon called *flux flow resistance* : if a current is passed through a type II conductor exposed to a magnetic field larger than B_{c1}, the current exerts a force on the magnetic flux lines which then begin to move through the material and generate frictional heat. So effectively a type II superconductor behaves like an Ohmic resistor although the current flow itself is resistance-free. In order to be useful for the application in magnet coils the flux motion must be prevented by capturing the flux lines at so-called *pinning centres*. Type II conductors with good pinning are called *hard superconductors* and these are indeed the materials useful for magnets. We will see that flux pinning is associated with an undesirable side-effect: the superconductor acquires a pronounced magnetic hysteresis which is responsible for the persistent-current field distortions.

2.2   Meissner-Ochsenfeld effect and London penetration depth

We consider a cylinder with perfect conductivity and raise a magnetic field from zero to a finite value B. A surface current is induced whose magnetic field, according to Lenz's rule, cancels the applied field in the interior. Since the resistance is assumed to vanish the current continues to flow with constant strength as long as the external field is kept constant, and consequently the bulk of the cylinder will stay field-free. This is exactly what happens if we expose a lead cylinder in the superconducting state $(T < T_c)$ to an increasing field, see the path $(a) \rightarrow (c)$ in Fig. 2.2. So below T_c lead acts as a perfect diamagnetic material. There is, however, another path leading to the point (c). We start with a lead cylinder in the normal state $(T > T_c)$ and expose it to a field which is increased from zero to B. Also in this case eddy currents are induced but they decay rapidly and after a few hundred microseconds the field lines will fully penetrate the material (state (b) in Fig. 2.2). Now the cylinder is cooled down. At the very instant the temperature drops below T_c a surface current is spontaneously created and the magnetic field is expelled from the interior of the cylinder. This surprising observation is called the *Meissner-Ochsenfeld effect* after its discoverers; as mentioned earlier it cannot be explained by the law of induction because the magnetic field is kept constant.

In a (T, B) plane, the superconducting phase is separated from the normal phase by the curve $B_c(T)$ as sketched in Fig. 2.3. Also indicated are the two ways on which one can reach the point (c). It is instructive to compare this with the response of a 'normal' metal of perfect conductivity. The field increase along the path $(a) \rightarrow (c)$ would yield the same result as for the superconductor, however the cooldown along the path $(b) \rightarrow (c)$ would have no effect at all. So superconductivity means definitely more than just vanishing resistance.

The first successful attempt to explain the Meissner-Ochsenfeld effect was undertaken in 1935 by Heinz and Fritz London. They assumed that the supercurrent is carried by a fraction of the conduction electrons in the metal. The 'super-electrons' experience no frictional force, so their equation of motion in an electric field is

$$m\frac{\partial \vec{v}}{\partial t} = -e\vec{E}$$

leading to an accelerated motion. The super-current density is

$$\vec{J_s} = -en_s\vec{v}$$

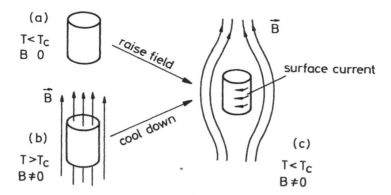

Figure 2.2: A lead cylinder in a magnetic field. Two possible ways are sketched to reach the superconducting state with $B > 0$.

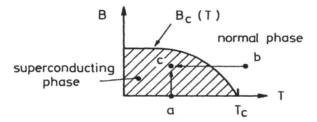

Figure 2.3: The phase diagram in a (T,B) plane with two ways to reach point (c) in the superconducting region.

where n_s is the density of the super-electrons. This yields the equation

$$\frac{\partial \vec{J}_s}{\partial t} = \frac{n_s e^2}{m} \vec{E} \ .$$ (2.3)

Now one uses the Maxwell equation

$$\vec{\nabla} \times \vec{E} = -\frac{\partial \vec{B}}{\partial t}$$

and inserts $\vec{E}$ from Eq. (2.3) to obtain

$$\frac{\partial}{\partial t} \left(\frac{m}{n_s e^2} \vec{\nabla} \times \vec{J}_s + \vec{B} \right) = 0 \ .$$

Since the time derivative vanishes the quantity in the brackets must be a constant. Up to this point the derivation has been fully compatible with classical electromagnetism, applied to the frictionless acceleration of electrons, for example to the motion in the vacuum of a television tube or a circular accelerator. The essential new assumption H. and F. London made is that the bracket is not an arbitrary constant but is equal to zero. Then one obtains the important *London equation*

$$\vec{\nabla} \times \vec{J}_s = -\frac{n_s e^2}{m} \vec{B} \; . \tag{2.4}$$

It should be noted that the above assumption cannot be justified within classical physics and is indeed wrong for normal conductors. Combining the fourth Maxwell equation (for time-independent fields)

$$\vec{\nabla} \times \vec{B} = \mu_0 \vec{J}_s$$

and the London equation and making use of the relation

$$\vec{\nabla} \times (\vec{\nabla} \times \vec{B}) = -\nabla^2 \vec{B}$$

(this is valid since $\vec{\nabla} \cdot \vec{B} = 0$) we get the following equation for the magnetic field in a superconductor

$$\nabla^2 \vec{B} - \frac{\mu_0 n_s e^2}{m} \vec{B} = 0 \; . \tag{2.5}$$

For a simple geometry, namely the boundary between a superconducting half space and vacuum, and with a magnetic field parallel to the surface, Eq. (2.5) reads

$$\frac{d^2 B_y}{dx^2} - \frac{1}{\lambda_L^2} B_y = 0 \; .$$

Here we have introduced a very important superconductor parameter, the *London penetration depth*

$$\lambda_L = \sqrt{\frac{m}{\mu_0 n_s e^2}} \quad . \tag{2.6}$$

The solution of the differential equation is

$$B_y(x) = B_0 \exp(-x/\lambda_L) \; .$$

So the magnetic field does not stop abruptly at the superconductor surface but penetrates into the material with exponential attenuation (Fig. 2.4a). For typical material parameters the penetration depth is quite small, namely 20 – 50 nm. In the bulk of a thick superconductor the magnetic field vanishes which is just the Meissner-Ochsenfeld effect. Here it is appropriate to recall that in the BCS theory not single electrons but pairs of electrons are the carriers of the supercurrent. The penetration depth, however, remains unchanged since going from single electrons to Cooper pairs implies the following replacements

$$m \rightarrow 2m, \; e \rightarrow 2e, \; n_s \rightarrow n_s/2 \; .$$

The superconductor can tolerate a magnetic field only in a thin surface layer. An immediate consequence is that current flow is restricted to the same layer. Currents in the interior are forbidden as they would generate magnetic fields in the bulk. If we pass a current through a lead wire it flows therefore in a surface sheath of about 20 nm thickness, see Fig. 2.4b, so the overall current in the wire is small. This is another reason, besides the low critical field, why type I superconductors are inadequate for winding superconducting magnet coils.

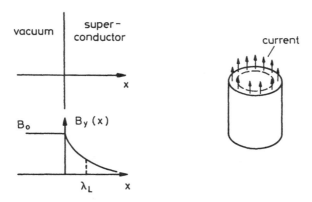

Figure 2.4: Exponential attenuation of a magnetic field in a superconducting half plane and current flow in a wire made from a type I superconductor.

2.3 Type II superconductors

2.3.1 Energy balance in a magnetic field

A material like lead goes from the normal to the superconducting state when it is cooled below T_c and when the magnetic field is less than $B_c(T)$. This is a phase transition comparable to the transition from water to ice below $0°$ C. Phase transitions take place when the new state is energetically favoured. The relevant thermodynamic energy is here the so-called Gibbs free energy G (see e.g. D.R. Tilley and J. Tilley (1990)). Free energies have been measured for a variety of materials. For temperatures $T < T_c$ they are found to be lower in the superconducting than in the normal state while G_{super} approaches G_{normal} in the limit $T \to T_c$, see Fig. 2.5a. What is now the impact of a magnetic field on the energy balance? A magnetic field has an energy density $B^2/2\mu_0$, and according to the Meissner-Ochsenfeld effect the magnetic energy must be pushed out of the lead cylinder when it enters the superconducting state. Hence the free energy in the superconducting state increases quadratically with the applied field:

$$G_{super}(B) = G_{super}(0) + \frac{B^2}{2\mu_0} \cdot V_m \qquad (2.7)$$

where V_m is the volume of a mole (10 cm^3 for Al). The normal-state energy remains unaffected by the magnetic field. The material stays superconductive as long as $G_{super}(B) < G_{normal}$. Equation (2.7) implies the existence of a maximum tolerable field, the 'critical field', above which superconductivity breaks down. It is defined by the condition that the free energies in the superconducting and in the normal state be equal

$$G_{super}(B_c) = G_{normal} \Rightarrow \frac{B_c^2}{2\mu_0} \cdot V_m = G_{normal} - G_{super}(0) . \qquad (2.8)$$

Figure 2.5b illustrates what we have said. For $B > B_c$ the normal phase has a lower energy, so the material goes to the normal state.

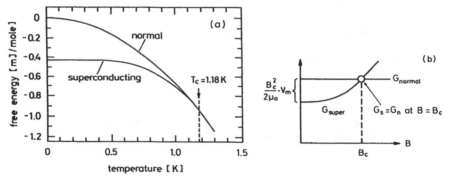

Figure 2.5: (a) The measured free energies of aluminium in the normal and the superconducting state as a function of temperature (after N.E. Phillips). The normal state was achieved by exposing the sample to a magnetic field larger than B_c. The free energy is essentially given by $G = U - T \cdot S$ where U is the internal energy and S the entropy. The second term dominates, hence G is negative. (b) Schematic sketch of the free energies as a function of magnetic field.

2.3.2 Coherence length and distinction between type I and type II superconductors

The above considerations have to be modified for very thin sheets of superconductor (thickness $< \lambda_L$). In the thin sheet the applied field does not drop to zero at the centre. Consequently less magnetic energy needs to be expelled which implies that the critical field of a thin sheet may be much larger than the B_c of a thick slab. From this point of view it might appear energetically favourable for a thick slab to subdivide itself into an alternating sequence of thin normal and superconducting slices. The magnetic energy is indeed lowered that way but there is another energy to be taken into consideration, namely the energy required to create the normal-superconductor interfaces. A subdivision is only sensible if the interface energy is less than the magnetic energy.

At the boundary between the normal and the superconducting phase the density n_c of the super-current carriers (the Cooper pairs) does not jump abruptly from zero to its value in the bulk but rises smoothly over a finite length ξ, called *coherence length*[2], see Fig. 2.6.

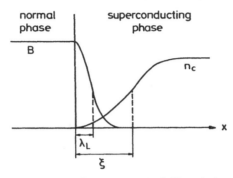

Figure 2.6: The exponential drop of the magnetic field and the rise of the Cooper-pair density at a boundary between a normal and a superconductor.

The relative size of the London penetration depth λ_L and the coherence length ξ decides whether a material is a type I or a type II superconductor. This we want to study in a semi-quantitative way. We define the *thermodynamical critical field* by the energy relation

$$\frac{B_{cth}^2}{2\mu_0} \cdot V_m = G_{normal} - G_{super}(0) \ . \tag{2.9}$$

For type I conductors, B_{cth} obviously coincides with B_c, see Eq. (2.8), while in the type II case B_{cth} lies between B_{c1} and B_{c2}. The difference between the two free energies, $G_{normal} - G_{super}(0)$, can be interpreted as the Cooper-pair condensation energy. For a conductor of unit area, exposed to a field $B = B_{cth}$ parallel to the surface, the energy balance is now as follows:
(a) The magnetic field penetrates a depth λ_L into the sample which corresponds to an energy gain since magnetic energy must not be driven out of this layer:

$$\Delta E_{magn} = \frac{B_{cth}^2}{2\mu_0} \cdot \lambda_L \ .$$

---

[2]More precisely this ξ is the Ginzburg-Landau coherence length ξ_{GL}, see e.g. W. Buckel (1990), D.R. Tilley and J. Tilley (1990).

(b) On the other hand, the fact that the Cooper-pair density does not assume its full value right at the surface but rises over a length ξ means a loss of condensation energy

$$\Delta E_{cond} = -\frac{B_{cth}^2}{2\mu_0} \cdot \xi \; .$$

Obviously there is a net energy gain if $\lambda_L > \xi$. So a subdivision of the superconductor into an alternating sequence of thin normal and superconducting slices is energetically favourable if the London penetration depth exceeds the coherence length.

A more refined treatment is provided by the Ginzburg-Landau theory (see e.g. D.R. Tilley and J. Tilley (1990)). Here one introduces the *Ginzburg-Landau parameter*

$$\kappa = \lambda_L/\xi \; . \tag{2.10}$$

The criterion for type I or II superconductivity is found to be

$$\text{type I:} \quad \kappa < 1/\sqrt{2}$$
$$\text{type II:} \quad \kappa > 1/\sqrt{2}.$$

The following table lists the penetration depths and coherence lengths of some important superconducting elements. Niobium is a type II conductor but close to the border to type I, while indium, lead and tin are clearly in the type I class.

material	In	Pb	Sn	Nb
λ_L [nm]	24	32	≈ 30	32
ξ [nm]	360	510	≈ 170	39

The coherence length ξ is proportional to the mean free path of the conduction electrons in the metal. In alloys the mean free path is generally much shorter than in pure metals so they are always type II conductors.

In reality a type II superconductor is not subdivided into thin slices but the field penetrates the sample in the form of flux tubes which arrange themselves in a triangular pattern which can be made visible with a beautiful technique developed by Essmann and Träuble (1967). The fluxoid pattern shown in Fig. 2.7a proves beyond any doubt that niobium is indeed a type II superconductor. Each flux tube or *fluxoid* contains one elementary flux quantum Φ_0 which is surrounded by a Cooper-pair vortex current. The centre of a fluxoid is normal-conducting. The area occupied by a flux tube is roughly $\pi\xi^2$. When we apply an external field B to a type II superconductor fluxoids keep moving into the specimen until their average field is identical to B. The fluxoid spacing in the triangular lattice is in this case

$$d = \sqrt{\frac{2\Phi_0}{\sqrt{3}B}}$$

which amounts to 20 nm at 6 Tesla. The upper critical field of a type II superconductor is reached when the current vortices of the fluxoids start touching each other at which point superconductivity breaks down. In the Ginzburg-Landau theory the upper critical field is given by

$$B_{c2} = \sqrt{2}\,\kappa\,B_{c\,th} = \frac{\Phi_0}{2\pi\xi^2}\;.\tag{2.11}$$

There exists no simple expression for the lower critical field. In the limit $\kappa \gg 1$ one gets

$$B_{c1} = \frac{1}{2\kappa}(\ln\kappa + 0.08)B_{c\,th}\;.\tag{2.12}$$

2.4 Hard superconductors

2.4.1 Flux flow resistance and flux pinning

For application in accelerator magnets a superconducting wire must be able to carry a large current in the presence of a field in the 5 – 10 Tesla range. Type I superconductors are definitely ruled out because their critical field is less than a few tenths of a Tesla and their current-carrying capacity is very small since the current is restricted to a thin surface layer (compare Fig. 2.4). Type II conductors appear promising at first sight: they feature large upper critical fields, and high currents are permitted to flow in the bulk material. However there is the problem of *flux flow resistance*. If a current flows through an ideal type II superconductor which is exposed to a magnetic field one observes heat generation. Figure 2.7b illustrates the mechanism. The current exerts a Lorentz force on the flux lines and causes them to move through the specimen in a direction perpendicular to the current and to the field. This is a viscous motion and leads to heat generation. So although the current itself flows without dissipation the sample acts as if it had an Ohmic resistance. Flux flow resistance has been experimentally established (Kim et al. 1965).

To obtain useful wires for magnet coils flux flow has to be prevented. The standard method is to capture the fluxoids at *pinning centres*. These are defects or impurities in the regular crystal lattice. The most important pinning centres in niobium-titanium are normal-conducting titanium precipitates in the so-called α phase whose size is in the range of the average fluxoid spacing, see Chap. 3. There is another way to look at current flow in the bulk of a type II superconductor. Take a superconducting slab whose surface is in the yz plane and let us assume that there exists a magnetic field B_y in y direction and a current density J_z in z direction. Then the Maxwell equation $\vec{\nabla} \times \vec{B} = \mu_0 \vec{J}$ reads

$$\frac{\partial B_y}{\partial x} = \mu_0 J_z$$

which implies that a non-vanishing current density inside the conductor is necessarily coupled with a gradient in magnetic flux density. Such a gradient can only be maintained if flux pinning exists.

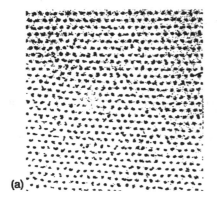 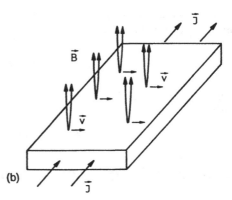

Figure 2.7: (a) Fluxoid pattern in niobium (courtesy U. Essmann). The distance between adjacent flux tubes is 0.2 μm. (b) Scheme of fluxoid motion in a current-carrying type II superconductor.

A type II superconductor with strong pinning is called a *hard superconductor*. Hard superconductors are very well suited for high-field magnets, they permit dissipationless current flow in high magnetic fields. There is a penalty, however: these conductors exhibit a strong magnetic hysteresis which is the origin of the very annoying 'persistent-current' multipoles in superconducting accelerator magnets.

2.4.2   Magnetization of a hard superconductor

A type I superconductor shows a completely reversible response to an external magnetic field B_e. The magnetization M is a unique function of the field[3], namely the straight line $M(B_e) = -B_e/\mu_0$ for $0 < B_e < B_c$. An ideal type II conductor without any flux pinning should also react reversibly. A hard superconductor, on the other hand, is only reversible in the Meissner phase because then no magnetic field enters the bulk, so no flux pinning can happen. If the field is raised beyond B_{c1} magnetic flux enters the sample and is captured at pinning centres. When the field is reduced again these flux lines remain bound and the specimen keeps a frozen-in magnetiza-

---

[3]There is often a confusion whether the magnetic field should be the H or the B field. D.R. Tilley and J. Tilly (1990) argue that the magnetization of a superconductor inside a current-carrying coil resembles that of an iron core. Then the 'magnetizing' field H is appropriate which is generated by the coil current only, and the total field in the material is given by the superposition of H and the superconductor magnetization $M = M(H)$: $B = \mu_0(H + M)$. Often, however, the external field is produced by an electromagnet with iron pole shoes, and this is already a B field. Since B appears also in the Lorentz force we prefer this quantity. With this convention, the field B_i inside a superconductor is given in terms of the external field B_e by $B_i = B_e + \mu_0 M$. Unfortunately much of the superconductivity literature is based on the obsolete CGS system where the distinction between B and H is not very clear and the two fields have the same dimension although their units were given different names: Gauss and Oerstedt.

tion even for vanishing external field. One has to invert the field polarity to achieve $M = 0$ but the initial state ($B_e = 0$ and no captured flux in the bulk material) can only be recovered by warming up the specimen to destroy superconductivity and release all pinned flux quanta, and by cooling down again. A typical hysteresis curve is shown in Fig. 2.8. There is a close resemblence with the hysteresis in iron except for the sign: the magnetization in a superconductor is opposed to the magnetizing field because the physical mechanism is diamagnetism.

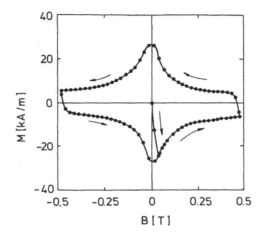

Figure 2.8: Measured magnetization M of a multifilamentary niobium-titanium conductor (Collings et al. 1990). Shown is the initial excitation, starting at $B = 0$ and $M = 0$, and the magnetic hysteresis for an external field B varying between +0.5 T and -0.5 T. Note that the hysteresis curve is not exactly symmetric with respect to the horizontal axis. The slight asymmetry is due to the magnetic moment generated by surface currents (Meissner-Ochsenfeld effect) which is always opposite in sign to the applied field.

The magnetic hysteresis is associated with energy dissipation. When a hard superconductor is exposed to a time-varying field and undergoes a cycle like the loop in Fig. 2.8, the energy loss is given by the integral

$$Q_{hyst} = \oint M(B)dB \ . \tag{2.13}$$

It is equal to the area enclosed by the loop. This energy must be provided by the power supply of the field-generating magnet. It is transformed into heat in the superconductor because frictional heat is produced when magnetic flux quanta are moved in and out of the specimen. This *hysteretic loss* must be taken into consideration in the heat load budget of the cryogenic plant of a superconducting accelerator. For an experimental determination see Sect. 7.2.

2.4.3   Flux creep

The pinning centres prevent flux flow in hard superconductors but some small *flux creep* effects remain. At finite temperatures, even as low as 4 K, a few of the flux quanta may be released from their pinning locations by thermal energy and move out of the specimen, thereby reducing the magnetization. The first flux creep experiment was carried out by Kim and co-workers (1962) with a small NbZr tube. If one plots the internal field at the centre of the tube as a function of the external field the well-known hysteresis curve is obtained in which one can distinguish the shielding and the trapping branch, see Fig. 2.9a. The Kim group observed that on the trapping branch the internal field exhibited a slow logarithmic decrease with time while on the shielding branch a similar increase was seen (Fig. 2.9b). Both observations can be explained by assuming a logarithmic time decay of the critical current density.

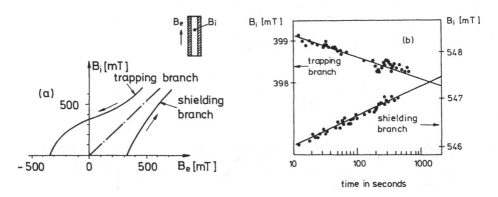

Figure 2.9: (a) Hysteresis of the internal field in a tube of hard superconductor. (b) Time dependence of the internal field on the trapping and the shielding branch.

A theoretical model for thermally activated flux creep was proposed by Anderson (1962). The pinning centres are represented by potential wells of average depth U_0 and width a in which bundles of flux quanta with an average flux $\langle \Phi \rangle$ are captured. At zero current the probability that magnetic flux leaves a potential well is proportional to the Boltzmann factor

$$P_0 \propto \exp(-U_0/k_B T) .$$

When the superconductor carries a current density J the potential acquires a slope proportional to the driving force density $F \propto \langle \Phi \rangle \cdot J$. This slope reduces the effective potential well depth to $U = U_0 - \Delta U$ with $\Delta U \approx \langle \Phi \rangle \cdot J a\, l_b$, see Fig. 2.10. Here l_b is the length of the flux bundle. The probability for flux escape increases to

$$P = P_0 \exp(+\Delta U/k_B T) .$$

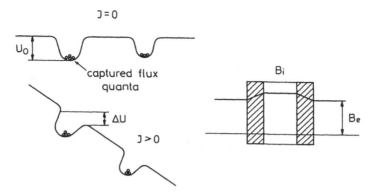

Figure 2.10: Sketch of the pinning potential without and with current flow and field profile across the NbZr tube.

We consider now the tube in the Kim experiment at a high external field B_e on the trapping branch of the hysteresis curve. The internal field is then slightly larger, namely by the amount $B_i - B_e = \mu_0 J_c w$ where J_c is the critical current density at the given temperature and magnetic field and w the wall thickness. Under the assumption $B_i - B_e \ll B_e$ both field and current density are almost constant throughout the wall. The reduction in well depth ΔU is proportional to the product of these quantities. If a bundle of flux quanta is released from its well, it will 'slide' down the slope and leave the material. In this way space is created for magnetic flux from the bore of the cylinder to migrate into the conductor and refill the well. As a consequence the internal field decreases and with it the critical current density in the wall. Its time derivative is roughly given by the expression

$$\frac{dJ_c}{dt} \approx -C \exp\left(\frac{\Delta U}{k_B T}\right) \approx -C \exp\left(\frac{\langle\Phi\rangle a J_c l_b}{k_B T}\right) \tag{2.14}$$

where C is a constant. The solution of this unusual differential equation is

$$J_c(t) = J_c(0) - \frac{k_B T}{\langle\Phi\rangle a\, l_b} \ln t \ . \tag{2.15}$$

This result implies that for given temperature and magnetic field the critical current density is not really a constant but depends slightly on time. What one usually quotes as J_c is the value obtained after the decay rate on a linear time scale has become unmeasurably small. If we differentiate $J_c(t)$ with respect to $\ln t$ we obtain the *logarithmic decay rate*

$$R_0 = \frac{dJ_c(t)}{d(\ln t)} = -\frac{k_B T}{\langle\Phi\rangle a\, l_b} \tag{2.16}$$

which is proportional to the temperature T.

2.4.4   Critical current density

For a hard superconductor, not only temperature T and magnetic field B have to be specified but also current density J. The material can be conveniently characterized by its *critical surface* in a (T, B, J) coordinate system. For the most important conductor used in magnets, niobium-titanium, this surface is depicted in Fig. 2.11. Superconductivity prevails everywhere below the surface and normal conductivity above it. Due to the presence of flux creep a hard superconductor is not exactly free of any resistance. The critical current density (at a given temperature and field) is usually defined by the criterion that the resistivity be $\rho = \rho_c = 10^{-14}\,\Omega\mathrm{m}$. In the vicinity of this point the resistivity is a very steep function of current density. It can be parametrised with an exponential or with a power law

$$\rho(J) = \rho_c \left(\frac{J}{J_c}\right)^n . \tag{2.17}$$

The quantity n is a quality index which may be as large as 50 for a good multifila-mentary NbTi conductor (see Chap. 3).

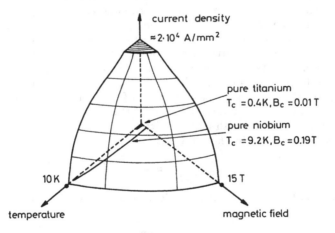

current density
$\approx 2 \cdot 10^4$ A/mm^2

pure titanium
$T_c = 0.4\,\mathrm{K}, B_c = 0.01\,\mathrm{T}$

pure niobium
$T_c = 9.2\,\mathrm{K}, B_c = 0.19\,\mathrm{T}$

10 K

15 T

temperature

magnetic field

Figure 2.11: Sketch of the critical surface of NbTi. Also indicated are the regions where pure niobium and pure titanium are superconducting. The critical surface has been truncated in the regime of very low temperatures and fields where only sparse data are available.

The fact that the resistance does not vanish implies of course that a 'persistent' current in a hard superconductor should slowly decrease. Equation (2.17) can actually be used to derive the logarithmic time dependence that is typical for flux creep[4]. Let us consider the NbZr tube again. When a current density J is flowing in the material

---

[4] We thank S.L. Wipf for pointing this out to us.

there is a resistance

$$R = R_c \left(\frac{J}{J_c} \right)^n$$

with $R_c = \rho_c 2\pi r / (w\,h)$. Here r is the average radius of the tube, h its height and w the wall thickness. Let L be the inductance of the tube. The current density obeys the differential equation

$$L\frac{dJ}{dt} + RJ = 0$$

which for constant resistance would lead to the usual exponential decay, $J(t) = J_0 \exp(-(R/L)t)$. Here, however, we have to solve the equation

$$L\frac{dJ}{dt} + R_c J_c \left(\frac{J}{J_c} \right)^{n+1} = 0 \,. \tag{2.18}$$

The solution is

$$J(t) = J_c \left(n\frac{R_c}{L}t + b \right)^{-1/n} \tag{2.19}$$

where b is an integration constant. The nth root can be expressed in the form

$$\left(n\frac{R_c}{L}t + b \right)^{-1/n} = \exp\left[-\frac{1}{n}\ln\left(n\frac{R_c}{L}t + b \right) \right] \,.$$

Since $n \gg 1$, the exponent is small, so we get in first order

$$J(t) = J_c \left(1 - \frac{1}{n}\ln\left(n\frac{R_c}{L}t + b \right) \right)$$

which is very similar to the logarithmic time dependence derived above.

2.4.5 The critical-state model and flux jumping

Critical state model

Starting from the observation that the resistivity of a hard superconductor is almost a step function of current density, C.P. Bean (1962, 1964) proposed the so-called *critical state model*, according to which there are only two possible states for current flow in a hard superconductor: the current density is either zero or equals the critical density J_c. The Meissner phase and the associated surface currents are ignored. The critical state model has proved very successful in describing the magnetization of hard superconductors. To illustrate this model we consider an initially unmagnetized slab of superconductor that is exposed to a magnetic field parallel to its surface. When the external field is raised from zero to some small value B_e a bipolar current of density $\pm J_c$ is induced in the slab which penetrates to such a depth that the shielding field cancels the applied field in the centre region. In the region of current flow the magnetic field exhibits a linear rise in accordance with the Maxwell equation

$\vec{\nabla} \times \vec{B} = \mu_0 \vec{J_c}$. Current and field profile are sketched in Fig. 2.12a. As long as the external field is kept constant the current pattern will persist. When the external field is increased both current and field penetrate deeper into the slab until the centre is reached (Fig. 2.12b). The associated field may be called the *penetrating field B_p*, its numerical value depends of course on the critical current density and the thickness of the slab. Raising B_e beyond B_p leads to a non-vanishing field at the centre but eventually the critical current density will drop because it depends on magnetic field.

An interesting situation occurs when B_e is lowered again. A new bipolar current of opposite polarity is induced and the current-field pattern inside the slab assumes the complicated shape sketched in Fig. 2.12c. It is straightforward to derive a hysteresis curve from this model. This shall not be done here as we will apply the critical-state model extensively in the computation of persistent-current field distortions in Chap. 6.

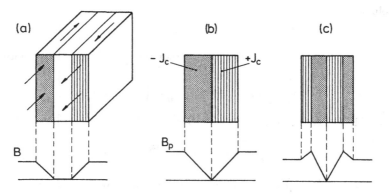

Figure 2.12: Current and field distribution in a slab of hard superconductor according to the critical-state model. The external field is parallel to the surface. (a) Initial exposition to a small external field. (b) The penetrating field B_p. (c) External field first raised above B_p and then lowered again.

The Bean model is based on the assumption of an infinitely steep transition from vanishing to finite resistance at $J = J_c$, corresponding to $n \to \infty$ in Eq. (2.17). L. Dresner (1995) has treated the more realistic case of a power-law transition with a typical n value of 50. He derives a diffusion-type differential equation for the current density. The sharp boundaries of the current flow region of Fig. 2.12a are smoothed out somewhat and move with an approximately logarithmic time dependence towards the centre of the slab. In Dresner's model the total current remains constant, hence the current density decreases inversely proportional to the width of the current-carrying zones. The time variation of superconductor magnetization is found to be rather similar to the logarithmic time dependence in our previous discussion which was based on a fixed penetration depth but a logarithmically decreasing critical current density. Since this latter case is easier to treat mathematically we stick to the original Bean model but allow for a time dependence of the critical current density.

Flux jumping

Early attempts to use hard superconductors for shielding or trapping of magnetic fields faced the difficulty that under certain conditions the supercurrents suddenly broke down. The underlying effect has become known under the name of *flux jumping*. Experimentally it was investigated by Lubell and Wipf (1965), a theoretical model for magnetic instabilities was presented by Swartz and Bean (1968). In this section we follow the simpler treatment in (Wilson 1983). Let us consider again the slab of superconductor but now exposed to an external field B_0 in y direction which is considerably larger than the penetrating field B_p. Choose the x axis as the normal of the slab and call $2a$ its thickness so that it extends from $x = -a$ to $x = +a$, see Fig. 2.13; the height h is assumed to much larger than the width $2a$. Applying the

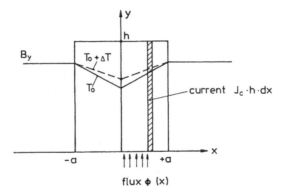

Figure 2.13: A slab of hard superconductor in a large external field parallel to its surface. Due to a heat input ΔQ the temperature rises from T_0 to $T_0 + \Delta T$ and the critical current density drops from J_c to $J_c - \Delta J_c$.

Maxwell equation

$$\oint \vec{B} \cdot d\vec{s} = \mu_0 I$$

(I is the current enclosed by the integration loop) one easily finds for the field inside the slab ($0 \le x \le a$)

$$B(x) = B_0 - \mu_0 J_c(a - x)h \ .$$

Here J_c is the critical current density at the given field B_0 and initial temperature T_0. Now assume that an amount of heat ΔQ per unit volume is put into the slab. The temperature will rise by a yet unknown amount ΔT and at the same time the critical current density will drop by ΔJ_c. What is the effect of reduced current density? The magnetic field inside the slab will increase, and the resulting change in magnetic flux is associated with a longitudinal voltage. This voltage leads to Joule heat generation because a current with density J_c is flowing in the slab. The time integral of the voltage is equal to the change in magnetic flux through the slab. We compute this for the half slab $0 \leq x \leq a$. The magnetic flux between 0 and x is $\phi(x) = \int_0^x B(x')dx' = B_0 x - \mu_0 J_c (ax - x^2/2)h$, and its change due to the change in J_c is

$$\Delta\phi(x) = \mu_0 \Delta J_c (ax - x^2/2)h \ .$$

The Joule heat produced in a slice of thickness dx and height h is $\Delta\phi(x)J_c\,dxh$. Integrating over x and dividing by the volume ah of the half-slab we obtain the Joule heat generation per unit volume

$$\Delta g = J_c \int_0^a \Delta\phi(x)dx = \mu_0 J_c \Delta J_c\, a^2/3 \ .$$

In first order one can write for the reduction in critical current: $\Delta J_c = J_c \cdot \Delta T/(T_c - T_0)$. Then the total energy balance reads

$$\Delta Q + \Delta g = C\Delta T \ \Rightarrow \ \Delta Q = [C - \mu_0 J_c^2 a^2/(3(T_c - T_0))]\Delta T \qquad (2.20)$$

where C is the specific heat per unit volume. From the second equation one realizes that the additional energy input Δg due to the Joule heating is equivalent to a reduction in heat capacity. We define an effective specific heat by

$$\tilde{C} = C - \mu_0 J_c^2 a^2/(3(T_c - T_0)) \ . \qquad (2.21)$$

Then it is obvious that an instability is reached when this quantity vanishes or becomes negative. In that case the slightest disturbance will cause the superconductor to reduce its critical current and expel part of the captured magnetic flux. This process is called flux jumping. It must not necessarily lead to a quench since the specific heat of the conductor increases with the third power of temperature.

To prevent flux jumps the half-thickness of the slab must fulfill the inequality

$$a < \sqrt{\frac{3C(T_c - T_0)}{\mu_0 J_c^2}} \ .$$

For a superconducting cylinder, exposed to a transverse field, the maximum tolerable radius is given by almost the same expression

$$r_{max} = \frac{\pi}{4}\sqrt{\frac{C(T_c - T_0)}{\mu_0 J_c^2}} \ . \qquad (2.22)$$

Let us apply this equation to niobium-titanium at $T_0 = 4.2$ K and $B_0 = 5$ T with $J_c \approx 3 \cdot 10^9$ A/m^2. From the formulae in Appendix D one gets $T_c(B_0) \approx 7.2$ K and $C \approx 5.6 \cdot 10^3$ J/(m^{3}K). The maximum permitted radius of the cylinder is computed to be $r_{max} = 30\,\mu$m. Hence wires made from pure superconductor are becoming instable against flux jumping if their diameter exceeds about 0.1 mm. This is the main motivation for using composite wires, made of many thin NbTi filaments which are embedded in a matrix of a high-conductivity normal metal, usually copper.

In the above considerations the cooling by the surrounding helium has been neglected. For this reason the relation (2.22) is called the adiabatic flux jump stability criterion.

References

P.W. Anderson, *Theory of flux creep in hard superconductor*, Phys. Rev. Lett. **9** (1962) 309

J. Bardeen, L.N. Cooper and J.R. Schrieffer, *Theory of superconductivity*, Phys. Rev. **108** (1957) 1175

C.P. Bean, *Magnetization of hard superconductors*, Phys. Rev. Lett. **8** (1962) 250

C.P. Bean, *Magnetization of high-field superconductors*, Rev. Mod. Phys. (1964) 31

W. Buckel, *Supraleitung*, 4. Auflage, VCH Verlagsgesellschaft, Weinheim 1990

E.W. Collings et al., *AC loss measurements of two multifilamentary NbTi composite strands*, Adv. Cryog. Eng. **36** (1990) 169

L. Dresner, *Stability of Superconductors*, Plenum Press, New York, London 1995

U. Essmann and H. Träuble, *The direct observation of individual flux lines in type II superconductors*, Phys. Lett. **24A** (1967) 526 and Sci. Am. **224**, March 1971

J. File and R.G. Mills, *Observation of persistent current in a superconducting solenoid*, Phys. Rev. Lett. **10** (1963) 93

Y.B. Kim, C.F. Hempstead and A.R. Strnad, *Critical persistent currents in hard superconductors* , Phys. Rev. Lett. **9** (1962) 306

Y.B. Kim, C.F. Hempstead and A.R. Strnad, *Flux flow resistance in type-II superconductors* , Phys. Rev. **139** (1965) A1163

N.E. Phillips, *Heat capacity of aluminum between 0.1 and 4.0 K*, Phys. Rev. **114** (1959) 676

P.S. Swartz and C.P. Bean, *A model for magnetic instabilities in hard superconductors: The adiabatic critical state*, J. Appl. Phys. **39** (1968) 4991

D.R. Tilley and J. Tilley, *Superfluidity and Superconductivity*, Third Edition, Institute of Physics Publishing Ltd, Bristol 1990

S.L. Wipf and M.S. Lubell, *Flux jumping in Nb-25%Zr under nearly adiabatic conditions*, Phys. Lett. **16** (1965) 103

Further reading

M.R. Beasley, R. Labusch and W.W. Webb, *Flux creep in type-II superconductors*, Phys. Rev. **181** (1969) 682

D.G. Cody, *Phenomena and Theory of Superconductivity*, in: H. Brechna, *Superconducting Magnet Systems*, Springer, Berlin 1973

R. Griessen, *Thermally activated flux motion near the absolute zero*, Physica **C172** (1991) 441

H. Ibach and H. Lüth, *Festkörperphysik*, 2. Auflage, Springer-Verlag, Berlin 1988

N.W. Ashcroft, N.D. Mermin, *Solid State Physics*, Saunders College Publishing, Fort
    Worth 1976
C. Kittel, *Introduction to Solid State Physics*, 6th edition, John Wiley, New York 1986
R.D. Parks (Ed.), *Superconductivity*, Dekker, New York 1969
M. Tinkham, *Introduction to Superconductivity*, Krieger Publ., Malabar, Florida 1980

Chapter 3

Practical Superconductors for Accelerator Magnets

3.1 Superconducting materials

A large variety of metals and alloys become superconductive at liquid helium temperatures but basically only two materials are commercially available for large scale magnet production, niobium-titanium NbTi and niobium-tin Nb₃Sn. Typical critical current densities achieved in recent conductors are plotted in Fig. 3.1a as a function of field.

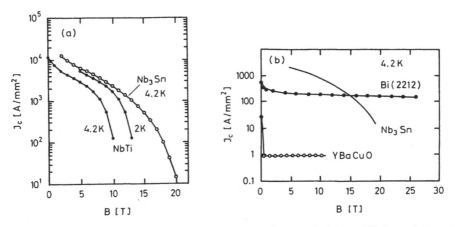

Figure 3.1: (a) Critical current density as a function of magnetic field in NbTi at 4.2 K and 2.0 K and in Nb_3Sn at 4.2 K (Krauth 1990). (b) Critical current density at 4.2 K of a wire made from $Bi_2Sr_2Ca_1Cu_2O_4$ in comparison with Nb_3Sn and $YBa_2Cu_3O_7$ (Tenbrink et al. 1991). (© 1991 IEEE)

The 'work horse' is still niobium-titanium, in spite of the fact that its upper

critical field is only 10 T at 4.2 K. The outstanding feature of NbTi is its extreme
ductility which permits effective and simple fabrication methods for wires and cables.
For this reason it is widely used in magnets of moderate field strength (up to 6.5 T
at 4.2 K). Cooling with superfluid helium of 2 K increases the field level to about 9
T. The critical temperature and upper critical field vary with composition as shown
in Fig. 3.2. The optimum titanium proportion in the alloy is 46.5 weight %.

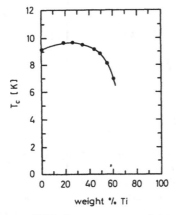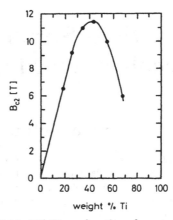

Figure 3.2: Critical temperature and upper critical field of NbTi as a function of composition
(Meingast et al. 1989). (© 1989 AIP)

For higher fields Nb_3Sn or $(Nb, Ta)_3Sn$ with upper critical fields of about 20 T
at 4.2 K are used. Niobium-tin is a brittle intermetallic compound of well-defined
stoichiometry which crystallizes in the so-called A15 lattice. Because of their brittleness the A15 superconductors cannot be drawn to thin filaments like NbTi but must
be formed in the final geometry by high-temperature heat treatment. Nb_3Sn, for
example, has to be 'reacted' at 650 to 700° C for many days to achieve full performance. In the fully reacted state multifilamentary Nb_3Sn strands are quite sensitive
to mechanical stress. Bending with a small radius of curvature leads to a severe loss
in current-carrying capacity. Accelerator dipoles or quadrupoles can therefore not be
wound from the fully reacted cable because a strong degradation would happen in the
coil head region. Rather the heat treatment must be performed after coil winding.
Bare wires and Rutherford cables with open voids are moreover strongly affected by
the transverse stress in the pre-compressed coil. The measured reduction of critical
current in a cable as a function of transverse stress is plotted in Fig. 3.3 (Pasztor
et al. 1994). A large improvement is gained when the voids in the cable are filled
with epoxy or solder because then the applied pressure is evenly distributed and local
stress enhancements are avoided. In an epoxy-filled cable the critical current drops
by only 5% for transverse stresses of up to 150 MPa. Data by Jakob et al. (1991)
lead to the same favourable result. Hence an epoxy-impregnated Nb_3Sn coil should

not suffer appreciably from superconductor degradation due to precompression of the coil and magnetic forces.

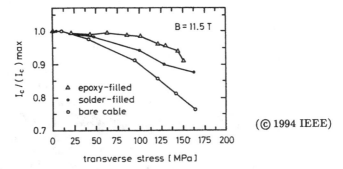

(© 1994 IEEE)

Figure 3.3: Degradation in critical current as a function of the applied stress for a bare Nb$_3$Sn cable and for cables filled with epoxy or solder. Filling the voids with epoxy or solder reduces local stress maxima and improves the performance (Pasztor et al. 1994).

Since the discovery of 'high-T_c' superconductivity in ceramic copper oxides considerable effort has been put into the development of tapes and wires for magnet coils. For temperatures below 30 K the ceramic superconductors appear promising. Fig. 3.1b shows that Bi$_2$Sr$_2$Ca$_1$Cu$_2$O$_4$ (BSCCO 2212) wires can carry reasonable current densities in fields of 25 Tesla or more which are inaccessible to low-T_c superconductors. So these materials open the way to superconducting magnets of very high field. At liquid nitrogen temperature, however, the application of wires has not proven useful so far because reasonable current densities could be achieved in low magnetic fields only, making the material useful for current leads (see e.g. Ballarino and Ijspeert 1994) but not for magnets. This is different for thin films. High critical current densities in the plane of the film have been measured at 77 K and in high magnetic fields.

3.2 Superconducting wire

3.2.1 Properties of multifilamentary NbTi wires

The critical parameters of a superconducting wire made from a hard superconductor have different origin (Wilson 1995): critical temperature and field are essentially determined by the chemical composition of the material while the critical current density is influenced by the microstructure and can be varied over a wide range, depending on the details of the production process.

The cables used in winding the coils of accelerator magnets consist of 20–40 wires (strands) of about 1 mm diameter. It is impossible to use wires made of pure superconductor for this purpose. They would be extremely vulnerable against flux jumps during excitation, the release of magnetic flux bundles from their pinning centres. Such flux jumps are accompanied with a heating of the material, often beyond the critical temperature. Since NbTi or other superconductors have a fairly high resistivity in the normal state, the consequence would be a quench of the coil caused by heating due to the transport current[1]. Equation (2.22) shows that only rather thin superconductor filaments (diameter below 100 μm for NbTi at 5 T and 4.2 K) are stable against flux jumps. In a multifilamentary wire a large number of thin filaments are embedded in a copper matrix that provides mechanical stability and at the same time serves as an electrical bypass of high conductivity and as a heat sink. Cross sections of two typical strands are shown in Fig. 3.4. If a filament should temporar-

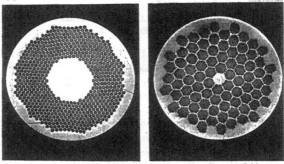

Figure 3.4: Cross sections of two NbTi multifilamentary strands (diameter 0.84 mm) made by Vacuumschmelze. Left side: strand used in the HERA quadrupoles, 636 filaments of 19 μm diameter. Right side: prototype strand for an LHC dipole with 10164 filaments of 5 μm diameter, made by a two-stage extrusion and stacking procedure. (Courtesy H. Krauth).

ily be heated beyond the critical temperature, for instance due to a small flux jump, the current is taken over by the copper for a short moment, allowing the NbTi to cool down and recover superconductivity. In order to fulfil these tasks the copper matrix must be in as good an electrical and thermal contact with the superconductor as possible. High purity copper with excellent electrical and thermal conductivity at 4.2 K is needed. The so-called *residual resistivity ratio* $RRR = R(300\,\text{K})/R(10\,\text{K})$ should exceed 100. Going to still larger RRR values is of not much use since copper exhibits a *magneto-resistance* which limits the low-temperature conductivity. The reason is that the Lorentz force changes the trajectories of the conduction electrons. The additional resistivity due to a transverse magnetic field is roughly given by the Köhler rule

$$\rho_m \approx a \cdot B \tag{3.1}$$

with $a = 4.5 \cdot 10^{-11} \Omega\text{m/T}$ for copper. In a field of 5 T the magneto-resistivity of $\rho_m \approx 2.2 \cdot 10^{-10}\,\Omega\,\text{m}$ corresponds to the low-temperature resistivity of copper with $RRR \approx 80$ in the absence of a magnetic field. Aluminium exhibits a different response

---

[1]Stability will be discussed in detail in Chap. 8.

to magnetic fields. Its magneto-resistance rises fast at low fields but saturates at a value that is an order of magnitude lower than in copper. Hence aluminium would be a far better stabilizer but co-extrusion of NbTi with aluminium has not proven practical[2].

Magnetic instabilities limit not only the thickness of the NbTi filaments but also the diameter of a multifilamentary wire. Here the instability derives from the *self-field* which is produced by the transport current in the wire. Since the self field starts to penetrate the wire from the outside the current distribution is not homogeneous throughout the cross section. At low excitation the transport current flows only in an outer shell of the wire but with increasing current it penetrates further towards the centre. The associated magnetic flux change leads to a longitudinal voltage (the 'flux-flow' voltage) which gives rise to dissipation. The condition for stability, which can be derived by similar arguments as used in Sect. 2.4.5, sets an upper limit for the wire diameter (Wilson 1983, ten Kate 1988)

$$d_{wire} < \sqrt{\frac{32C(T_c - T_0)}{\mu_0 J^2}} \Big/ \sqrt{-2\ln(1-i) - 2i - i^2} \,. \tag{3.2}$$

Here J is the average current density in the composite wire and $i = I_t/I_c$ the ratio of transport to critical current. At $B_0 = 5$ T and $J = 500$ A/mm^2 one finds a maximum permitted diameter of 2 mm. The relation (3.2) is called the *self-field stability criterion*.

Good flux pinning is the prerequisite for achieving high current densities in the presence of a large magnetic field. The most important pinning centres in niobium-titanium are normal-conducting titanium precipitates in the so-called α phase whose size is in the range of the fluxoid spacing (≈ 10 nm at 6 Tesla). Figure 3.5 shows a microscopic picture of a recent conductor with very high current density (3700 A/mm^2 at 5 T and 4.2 K). The optimization of the critical current density in NbTi involves a sequence of heat treatments, which precipitate the α-titanium at dislocation cell boundaries, and cold work, by which the wire is reduced to its final dimension (Meingast, Lee and Larbalestier 1989). The α-Ti precipitates are initially small islands of about 100 nm size. Very important for optimizing the current density is the final drawing strain. It is defined by the relation

$$\varepsilon_{fi} = 2\ln\left(d_i/d_f\right) \tag{3.3}$$

where d_i is the wire diameter at the last heat treatment and d_f the final diameter after drawing. During the final drawing step the α-Ti precipitates are deformed into thin ribbons whose thickness (1–2 nm) is less than the superconductor coherence length ($\xi \approx 5$ nm at 4.2 K). These ribbons are aligned along the wire axis and have a spacing of 3–6 nm (Meingast, Lee and Larbalestier 1989). Figure 3.6a shows that the

---

[2]In the large solenoid coils of storage ring experiments one often uses a Rutherford cable with copper matrix that is cladded with high-purity aluminium in a second extrusion step. Space limitations prevent the application of such well-stabilized conductors in accelerator magnets.

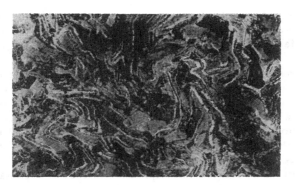

Figure 3.5: Micrograph of NbTi. The α-titanium precipitates appear as lighter strips. The area covered is 840 nm wide and 525 nm high. Courtesy P.J. Lee and D.C. Larbalestier.

critical current density rises with increasing drawing strain and reaches an optimum at $\varepsilon_{fi} = 5 - 6$.

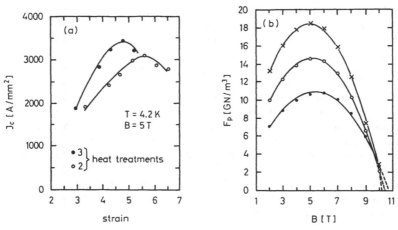

Figure 3.6: (a) Critical current density J_c as a function of final drawing strain ε_{fi}.
(b) Pinning force density at 4.2 K as a function of magnetic field for different heat treatments: $\times$ 80 h at 420°C, o 40 h at 375°C, • 5 h at 405°C (Li Chengren and Larbalestier 1987). (© 1987 Butterworth, Heinemann)

The pinning force density is given by the equation

$$\vec{F}_p = \vec{J}_c \times \vec{B} . \tag{3.4}$$

Figure 3.6b shows a typical plot of pinning force as a function of magnetic field. One observes a maximum at some intermediate level (5–7 T).

An important constraint on the filament diameter derives from the tolerable field distortions due to persistent magnetization currents. The magnetization is proportional to the product of filament diameter and critical current density, see Eq. (6.3), so the filament diameter should be minimized. This has become even more important in view of the significant increase in current density accomplished in recent years. With present technology, NbTi filament diameters of 5–6 μm are an optimum. A further reduction is costly and will eventually lead into the wrong direction because at very small inter-filament spacings a *proximity coupling* occurs, basically quantum-mechanical tunnelling of Cooper pairs through the normal material between adjacent filaments. In the coupling regime the superconductor magnetization grows with decreasing filament diameter, see Fig. 3.7a. A resistive matrix material such as CuMn

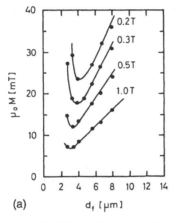
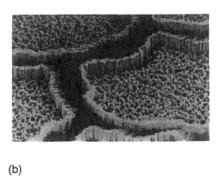

(a)   (b)

Figure 3.7: (a) Measured superconductor magnetization at various fields as a function of filament diameter d_f. The rise below 4 μm is caused by proximity coupling (Ghosh et al. 1987). (b) Scanning electron micrograph of 6 μm-diameter NbTi filaments surrounded by niobium diffusion barriers. The copper between the filaments has been etched away (courtesy A.F. Greene). (a: © 1987 IEEE)

is effective in suppressing interfilament coupling. If the copper matrix is alloyed with about 0.5 weight % of Mn, proximity coupling is barely perceptible even with filaments as thin as 1 μm (Collings, Marken and Sumption 1990). Another method of reducing the overall strand magnetization is based on ferromagnetic shielding either by replacing a few NbTi filaments with nickel filaments or by plating the strands with NiCu (Collings et al. 1991, Sumption and Collings 1992).

The field distortions from persistent currents are principally unavoidable and require a compensation by correction magnets. To facilitate these corrections, the three important parameters: filament diameter, critical current density and copper-to-superconductor (Cu/SC) ratio should be subjected to a tight quality control during wire and cable production. In manufacturing the HERA dipole cable, Brown Bovery was able to reduce the critical current variation to $\Delta I_c/I_c \leq 1\%$ by selecting the strands going into a cable according to their measured critical currents. The Cu/SC ratio was controlled to within $\pm 4\%$. The variation in magnetization $\Delta M/M$ was less than 2%, a factor of 2–3 smaller of what would have been obtained without these precautions (Maix et al. 1989).

The filaments in a strand are twisted with a pitch of typically 25 mm to suppress the eddy currents that are induced between different filaments during a field sweep (see Chap. 7). Also the strands in the cable are transposed with a pitch length of about 100 mm. In magnets for a.c. operation at 50 or 60 Hz one has to use very fine filaments (down to diameters of 0.5 μm) that are embedded in a CuNi matrix. A twist pitch of less than 2 mm is needed.

The ramp rate in proton storage rings is small, for instance 10 A/s (about 0.01 T/s) in HERA. In this case it is unnecessary to insulate the strands in the cable against each other, they may even be tinned in order to obtain a surface layer that prevents corrosion and simplifies soldering. For application at higher ramp rates, for instance 400 A/s in the Tevatron, all or at least every second strand are covered with an oxide layer in order to reduce cable eddy currents to an acceptable level.

3.2.2 Wire fabrication and test

Niobium-titanium

Superconductiong wires are fabricated in a multistep process. In our desription we follow Greene (1992) and concentrate on NbTi. The alloy NbTi must be produced with high purity and with a variation in titanium contents of 1% or less. From this so-called 'high homogeneity' material round bars are manufactured with a diameter of about 150 mm and a length of 500 to 750 mm. They are wrapped with a niobium foil of controlled tensile properties and grain size and inserted in a thick-walled can of pure copper (residual resistivity ratio $RRR = R(300\,\text{K})/R(10\,\text{K}) > 100$) of about 200 mm outer diameter. The can is closed at the ends by caps that are electron-beam welded. After evacuation the can is compressed and extruded at 600 to 700°C to a composite of 30–50 mm diameter. Now a multiple process of drawing and compaction is performed leading to a long bar of hexagonal cross section with a width of about 3.5 mm. This bar is cut into pieces of 0.50 to 0.75 m length. Several thousand of the carefully cleaned short bars are stacked into another 300 mm diameter thick-walled copper tube around a centre copper rod. The copper lids are electron-beam welded. Then again compaction, hot extrusion and a number of drawing steps follow with heat treatments in between to optimize the current-carrying capacity. Before the final drawing step the wire is twisted. If required, a tinning with final shaping or an anodizing is performed as the last step in wire production.

The niobium foil around the NbTi filament serves as a diffusion barrier (see Fig. 3.7b) to prevent the formation of CuTi compounds during the high-temperature heat treatment. These intermetallic compounds are brittle and do not reduce in size during the drawing procedure but fracture into small particles which might damage the filaments. If one aims for very thin filaments a second or even a third multifilamentary billet has to be produced from the hexagonal rods of an intermediate stage. The final wire may contain up to 100,000 filaments of about 1 μm diameter.

The wire production aims at manufacturing continuous lengths of several tens of kilometres with good and reproducible critical current characteristics. The wire thickness must be kept within a tolerance of a few μm. Inclusion of metallic chips or dirt during production must be strictly avoided as it may lead to wire breakage or reduced performance. Such foreign particles can be detected on-line by inductive methods. Samples from the ends of a continuous length are used for detailed investigation. Removing the copper reveals broken filaments. Scanning electron micrograph pictures are used to control the smoothness of the filament surface and the uniformity of the diameter. The critical current of the wires is routinely measured at 4.2 K at several field values taking at least two samples from each billet. A typical voltage-current curve is shown in Fig. 3.8a. The computation of the critical current density J_c in the superconductor itself requires the knowledge of the fractional area occupied by the NbTi. The Cu/SC ratio is determined by weighing a wire sample before and after etching off the copper or from a microscopic scan of the wire cross section. Critical current densities of $\approx$3100 A/mm^2 at 4.2 K and 5 T have been achieved in mass production of NbTi, a more than 50% increase since 1981 due to the use of improved material, Nb diffusion barriers and refined heat treatment.

As mentioned in Sect. 2.4.4 the transition to the normal state is not abrupt in a hard superconductor but the material becomes gradually resistive with a resistivity that is steeply rising with current. A frequently used parametrization is $\rho(J) \propto J^n$. The resistive transition index n is generally much larger than 1 and is a sensitive measure for the quality of the wire. Good wires may have n values as large as 50; if filaments are broken or vary in diameter along the length ('sausaging') one observes a reduction in n. For undamaged wires the quality index n depends linearly on the ratio B/B_{c2}, see Fig. 3.8b. It is easy to understand that n drops with increasing B since at higher fields not only the strongest but also weaker pinning centres are occupied. The strands of the HERA dipoles had $n \approx 40$ before and $n \approx 20$ after cabling indicating some filament damage during the cabling procedure (Wipf 1990).

Besides the superconductor properties a number of other aspects must be taken into account in strand design. In the cabling process the wire has to be flexible and bendable around small corners without degradation, that means without damaging the superconductor. The multifilamentary structure of NbTi in a copper matrix and the central copper core of the strand facilitate the deformation needed in cable production.

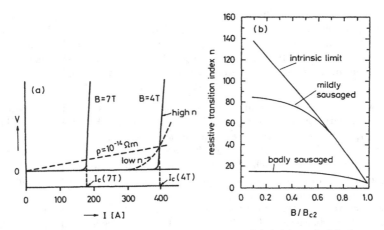

Figure 3.8: (a) Superconducting-to-normal transition and definition of critical current. The
$\rho = 10^{-14}\,\Omega$m criterion is indicated (ten Kate 1988). (b) Resistive transition index n as a
function of magnetic field (Warnes and Larbalestier 1986).   (© 1986 Butterworth,
Heinemann)

Niobium-tin

Several different methods exist for the production of Nb_3Sn wires (Thöner 1991,
Wilson 1995). In the bronze process filaments of pure niobium are drawn to their
final size in a matrix of CuSn bronze. In the following heat treatment the tin diffuses
through the bronze and reacts with the niobium to form Nb_3Sn. The electrical and
thermal conductivity of the bronze matrix is too low for a good stabilization of the
conductor, hence additional pure copper is needed for that purpose. The bronze
process is well suited for the production of fine filaments (typical diameter 2.5 μm).
An interesting alternative has been developed by ECN in Holland (Hornsveld et al.
1988). Here one starts with a copper matrix containing niobium tubes which are filled
with Nb_2Sn powder. During the heat treatment a Nb_3Sn reaction layer is formed but
the high conductivity of the copper matrix is preserved. The only drawback of this
high-performance conductor is the rather large effective filament diameter of 15 μm or
more. The ECN conductor was successfully used in the recent 11 T dipole prototype
(den Ouden et al. 1994, 1995) mentioned in Chap. 1.

Common to these and other production methods is the brittleness of the Nb_3Sn
conductor and its vulnerability to critical current degradation by strain.

3.3 Cable

Most superconducting accelerator magnets are made from a multi-strand cable of the so-called Rutherford type, sketched in Fig. 3.9. The wires are twisted and compressed into a flat two-layer cable. The cable is usually permeable to liquid helium so the surface of all strands is wetted with the coolant. Due to its high heat capacity the helium acts as a heat sink and stabilizes the conductor in case of magnetic flux jumps or transient heat production caused by wire motion. In Chap. 8 we will see that this is of utmost importance for good stability and quench performance of the magnets.

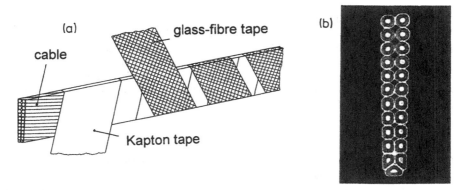

Figure 3.9: (a) Rutherford cable with Kapton and glass-fibre epoxy insulation such as used in the Tevatron and HERA magnets. For the RHIC and LHC coils an all-Kapton insulation is employed. (b) Cross section of the cable.

Ideally the cable should have a trapezoidal cross section that matches an appropriate azimuthal subdivision of the current shells in the magnet. This is technically possible for magnets with an inner diameter of 75 mm or more (Tevatron, HERA, RHIC) but meets difficulties at smaller diameters (SSC, LHC) since the compaction ratio $f_c = t_1/2d_s$ at the inner edge of the cable should stay above a tolerable lower limit of about 0.75. Here t_1 is the cable thickness at the narrow edge and d_s the strand diameter. Important cable parameters are the filling factor η_f, the trapezoid angle α_t and the compaction ratio f_c. The filling factor is:

$$\eta_f = \frac{N_s \pi (d_s/2)^2}{A_c \cos \beta_t} \tag{3.5}$$

with N_s = number of strands in the cable,

d_s = diameter of strand,

β_t = twist angle of cable,

A_c = cross sectional area of cable.

The cable design has to ensure tight packing of the strands to prevent wire motion during excitation. The number of strands ranges from 23 in the HERA quadrupoles to 36 in the outer layers of the SSC and LHC dipole coils. Too large a trapezoid angle bears the risk of filament damage at the strongly compressed narrow edge of the cable. The recent magnet designs (RHIC, SSC and LHC) are based on very moderate trapezoid angles of about 1.0°. However, angles as large as 4.6° have been realized at the KEK laboratory in Japan without much degradation (Hirabayashi et al. 1991). The cable corners must be round in order not to damage the insulation. The parameters of the HERA cable are:

$$\eta_f = 0.93, \ \alpha_t = 2.33°, \ f_c = 0.76 \ .$$

The residual resistivity ratio (RRR) of the strands is reduced in the cabling procedure due to the cold work the copper is subjected to. For the HERA dipole cable the reduction was from 100 to about 70. Most of this loss is recuperated when after coil winding the epoxy in the insulation is cured for several hours at temperatures around 150°C . The heat treatment is accompanied with a certain annealing of the copper, and in the case of SSC dipoles the residual resistivity ratio RRR was measured to rise again to values of 100 or more.

3.3.1   Cable fabrication

Cables are produced with machines equipped with the necessary number of wire spools. The wires are guided around a conical mandrel and then rolled to the required cross section by an assembly of rollers, named 'Turk's Heads'. The trapezoidal shape is achieved either in these rollers or in a second stage of rolling. Care must be taken to avoid breakage of strands as well as burrs and sharp edges on the cable surface which may puncture the cable insulation and lead to electrical shorts in the coil. The most endangered region is the narrow edge of the cable where the wires are strongly compressed and at the same time forced into a small radius of curvature. In some cases a final shaping process may turn out to be necessary. Experience shows that all components of the cabling machine must be well adjusted and the forces must be carefully controlled in order to avoid cable damage. Within any subcoil of a magnet all strands should be of continuous length without internal welds. While cold-welding of strands has often been used for economical reasons in case of breakage in the wire production, these non-superconducting transitions have a rather negative impact on the maximum current-carrying capacity of the coil and moreover they have been found to influence the time dependence of the superconductor magnetization (Ghosh et al. 1992).

A careful inspection of the finished cable is advisable. Inductive methods for on-line detection of broken wires have been successfully applied. Dimensional checks are performed right after the cabling machine. Devices have been invented that periodically clamp the cable with a preset compression, for instance 35 MPa, and measure the width, average thickness and trapezoid angle. A simple method to check the dimensions of the cable is to cut some 10 cable samples, stack them in alternating order in a fixture so that their wedge angles compensate and compress the assembly with the force they would experience in the collared coil. The size of the compressed cable stack allows to determine the cable mid thickness. By varying the compressive force the elastic modulus can be measured as well.

The degradation in critical current due to the cabling procedure is determined by testing cable samples at 4.2 K in an external magnetic field. These involved tests are usually not possible at the superconductor companies. The critical current is different for external fields perpendicular or parallel to the flat face of the cable, mainly because of the non-negligible self field of the cable. The results have to be corrected for this effect. The cable fabrication is considered satisfactory if the degradation in critical current is less than 2%.

The high precision needed in the coil dimensions puts stringent requirements on the tolerances of the cable. When the HERA conductor was specified in 1983 a tolerance interval of $\pm 20\,\mu$m on the cable width and thickness had to be granted although from a field quality point of view a much narrower tolerance on the cable thickness would have been highly desirable. In the course of the SSC and RHIC magnet program considerable effort has gone into the development of a much improved cabling machine. With present technology tolerances in the order of a few micrometres are accessible.

3.3.2   Cable insulation

The requirements on the cable insulation are demanding

- good electrical properties (breakdown voltage of more than 1000 V rms at 50–60 Hz in dry air)

- good mechanical properties (elasticity, yield strength) both at room and liquid helium temperature

- radiation hardness (the estimated lifetime dose for HERA is $\sim 5 \cdot 10^6$ Gray).

Most conventional insulating materials are excluded either by the low operating temperature or by the high radiation dose. One of the standard insulators at 4 K, Teflon, is not suited due to insufficient radiation hardness. The most commonly used materials are polyimide tape (e.g. Kapton) and glass-fibre tape with epoxy impregnation. The insulation of the Rutherford cable of the Tevatron and HERA magnets is shown in Fig. 3.9a. It consists of a double layer of Kapton, made by wrapping a 12 mm wide and 25 μm thick Kapton tape around the cable with an overlap of 58%, and a 9 mm wide and 120 μm thick glass-fibre tape. The glass tape is pre-impregnated with epoxy in the so-called 'B stage'; the polymerization of the epoxy starts only when it is cured at around 160°C. The epoxy content is 19 ± 2 weight % (Wolff 1985, 1987).

In the glass-tape layer there are gaps of 3 mm to allow liquid helium to penetrate the cable.

Kapton is a material with excellent electrical and satisfactory mechanical properties at cryogenic temperatures, see Appendix F. At room temperature some problems may arise due to the large internal pressure in the coil since Kapton starts to yield above 70 MPa. Moreover the foil is quite sensitive to burrs and sharp edges. The epoxy-impregnated glass-fibre tape serves basically as a protective layer of the Kapton insulation. It has the additional advantage of being somewhat compressible when the epoxy is heated in the curing press so that small variations in cable thickness are compensated. The compressibility can be controlled by adjusting the width of the gap in the glass tape layer. After polymerisation the glass-epoxy layer is no longer compressible. The unconventionally low epoxy contents (19%) of the glass tape is necessary to prevent the voids inside the cable from being filled with epoxy. As mentioned before, the aim is to have every strand surrounded with liquid helium for optimum cooling. Too low an epoxy content, on the other hand, leads to a coil whose windings are not rigidly glued together but may even fall apart.

The cable insulation in the RHIC magnets consists of a double wrap of Kapton (type CI) with a polyimide adhesive on the outside of the outer wrap. The coils are more accurate in azimuthal size than with glass-fibre wrapping (Wanderer et al. 1995). This type of nearly imcompressible insulation calls for tighter dimensional tolerances of the cable. Owing to the higher prestress needed in the coils of 8.4 T magnets, the inner polyimide insulation has been doubled for the LHC cable: a first layer is wound from 25 μm thick polyimide tape with 48% overlap, followed by a second layer with the same overlap which is wound in the opposite sense to avoid unwrapping in the coil head region. Finally a 70 μm thick polyimide tape is wrapped around the insulated cable, using the same winding sense as the first layer. This tape is covered with adhesive on the outside and 2 mm wide gaps are left between adjacent turns.

Before insulation the cable is cleaned in an ultrasonic bath in order to remove grease and dirt. When the Kapton wrapping is finished the cable is locally pressed with rollers and a high voltage insulation test is performed (e.g. 1 kV at 50–60 Hz).

Coils made from Nb_3Sn are manufactured in quite a different way because of the brittleness of the material. The 'react and wind' technique (the coil is wound from the fully reacted wire) may lead to a severe degradation of the superconductor during the winding process. The alternative 'wind and react' technique (the unreacted, flexible wire is wound and the coil is 'reacted' as a whole) has been successfully applied but requires a time-consuming heat treatment of the finished coil. The high firing temperature of 650 to 700°C is prohibitive for all organic insulating materials such as Kapton and allows only insulation materials like glass-mica tape (Cablosam or Suritex[3]) that can stand temperatures of 1000° C and are very radiation resistant.

---

[3]Cablosam is a registered trade mark of ISOLA, Switzerland, Suritex is a registered trade mark of ISOVOLTA, Austria.

After the firing procedure the coil is mechanically rather fragile. For the LHC project a 1-m-long prototype dipole was built by the ELIN company in Weiz, Austria (Asner et al. 1989). After heat-treatment it was vacuum-impregnated with epoxy to improve the mechanical stability. A similar procedure was applied for a short model magnet recently built by the University of Twente in Holland and Dutch industry (den Ouden et al. 1993, 1995). Of course the permeability for liquid helium is lost by the epoxy filling. The drawbacks in conjunction with the high costs have so far prevented the application of Nb_3Sn superconductor in the mass production of accelerator magnets.

References

A. Asner et al., *First Nb_3Sn, 1m long superconducting dipole model magnets for LHC break the 10 T threshhold*, Proc. 8th Int. Conf. on Magn. Techn. MT-8, Tsukuba 1989, p. 36

A. Ballarino and A. Ijspeert, *Expected advantages and disadvantages of high-T_c current leads for the Large Hadron Collider orbit correctors*, Adv. Cryog. Eng. **39** (1994) 1959

E.W. Collings et al., *Design, fabrication and properties of magnetically compensated SSC strands*, IEEE Trans. **MAG-27** (1991), 1787

E.W. Collings, K.R. Marken, M.D. Sumption, *Design of multifilamentary strand for Superconducting Supercollider (SSC) applications – reduction of magnetizations due to proximity effect and persistent current*, Supercollider 2, M. McAshan (ed.), Plenum Press, New York 1990, p. 581

A.K. Ghosh, W.B. Sampson, E. Gregory, T.S. Kreilick, *Anomalous low field magnetization in fine filament NbTi conductors*, IEEE Trans. **MAG-23** (1987) 1724

A.K. Ghosh, K.E. Robins and W.B. Sampson, *Time dependent magnetization effects in superconducting accelerator magnets*, Proc. XVth Int. Conf. on High Energy Accel., Hamburg 1992, p. 665

A. F. Greene, *Recent status of superconductors for accelerator magnets*, Proc. Workshop on ac Superconductivity, Tsukuba, KEK Proceedings 92-14 (1992) 100

H. Hirabayashi et al., *Design study of a superconducting dipole model magnet for the Large Hadron Collider*, IEEE Trans. **MAG-27** (1991) 2004

E.M. Hornsveld et al., *Development of ECN type niobium-tin wire towards smaller filament size*, Adv. Cryog. Eng. **34** (1988) 493

B. Jakob et al., *Reduced sensitivity of Nb_3Sn epoxy-impregnated cable to transverse stress*, Cryogenics **31** (1991) 390

H.H.J. ten Kate, *Practical Superconductors*, lectures at the CERN-DESY school 'Superconductivity in Particle Accelerators', Hamburg 1988, CERN report 89-04 (1989)

H. Krauth, *Development and large scale production of NbTi and Nb3Sn conductors for beam line and detector magnets*, Supercollider 2, M. McAshan (ed.), Plenum Press, New York 1990, p.631

Li Chengren, D.C. Larbalestier, *Development of high critical current densities in niobium 46.4 wt% titanium*, Cryogenics **27** (1987) 171

R.K. Maix, D. Salathe, S.L. Wipf, M. Garber, *Manufacture and testing of 465 km superconducting cable for the HERA dipole magnets*, IEEE Trans. **MAG-25** (1989) 1656

C. Meingast, P.J. Lee and D.C. Larbalestier, *Quantitative description of a high J_c Nb-Ti superconductor during its final optimization strain: I. Microstructure, T_c H_{c2} and resistivity*, J. Appl. Phys. **66** (1989) 5962

A. den Ouden et al., *An experimental 11.5 T Nb_3Sn LHC type of dipole magnet*, IEEE Trans. **MAG-30** (1994) 2320, and: *The Nb_3Sn dipole project at the University of Twente*, internal report 1995

G. Pasztor et al., *Transverse stress effects in Nb3Sn cables*, IEEE Trans. **MAG-30** (1994) 1938

M.D. Sumption and E.W. Collings, *Innovative strand design for accelerator magnets*, XVth Int. Conf. on High Energy Accel., Hamburg 1992, World Scientific 1993, p. 662

J. Tenbrink et al., *Development of high-Tc superconductor wires for magnet application*, IEEE Trans. **MAG-27** (1991) 1239

M. Thöner et al., *Nb_3Sn multifilamentary superconductors: an updated comparison of different manufacturing routes*, IEEE Trans. **MAG-27** (1991) 2027

P. Wanderer et al., *Construction and testing of arc dipoles and quadrupoles for the Relativistic Heavy Ion Collider (RHIC) at BNL*, Proc. Int. Conf. on High Energy Accel., Dallas 1995

W.H. Warnes, D.C. Larbalestier, *Critical current distributions in superconducting composites*, Cryogenics **26** (1986) 643

M.N. Wilson, *Superconducting materials for magnets*, Lectures at the CERN-DESY school 'Superconductivity at Particle Accelerators', Hamburg 1995, to be published

M.N. Wilson, *Superconducting Magnets*, Clarendon Press, Oxford 1983

S.L. Wipf, *Superconducting cable for HERA*, Supercollider 2, M. McAshan (ed.), Plenum Press, New York 1990, p. 557

S. Wolff, *The superconducting collared coil for dipoles of the proton ring of HERA - description and fabrication procedure*, DESY internal note, Feb. 1985, revised Feb. 1987

Further reading

E.W. Collings, *Applied Superconductivity, Metallurgy and Physics of Titanium Alloys*, Vol. 1: Fundamentals, Vol. 2: Applications, Plenum Press New York, London 1986

P. Dahl, *Oxide superconductors and the SSC*, internal report SSC-223 (1989)

S. Foner and B.B. Schwartz (Eds.), *Superconductor Material Science, Metallurgy, Fabrication and Applications*, Plenum Press New York, London 1981

A.K. Ghosh et al., *The effect of magnetic impurities and barriers on the magnetization and critical current of fine filament NbTi composites*, IEEE Trans. **MAG-24** (1988) 1145

C. Meingast and D.C. Larbalestier, *Quantitative description of a high J_c Nb-Ti superconductor during its final optimization strain. II. Flux pinning mechanisms*, J. Appl. Phys. **66** (1989) 5971

D.A. Pollock, *Quality analysis of superconducting wire and cable for SSC dipole magnets*, IEEE Trans. **MAG-28** (1991) 516

Chapter 4

Field Calculations

4.1 Multipole expansion for a single current conductor

A schematic view of a superconducting dipole for a large accelerator is given in Fig. 4.1. The length of the magnet is much larger than its aperture and the current

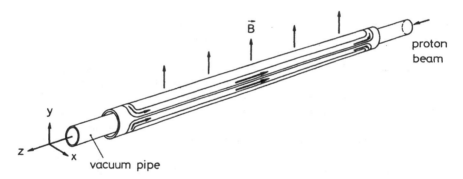

Figure 4.1: Schematic view of a superconducting dipole coil.

conductors run parallel to the beam over the longest part of the magnet, except for the short coil heads. Although the dipole magnets are usually not straight but follow the beam orbit the deviation from a straight line is only 18 mm for the 9-m-long HERA dipole. Under these circumstances one can consider the magnetic field essentially as two-dimensional and apply the theory of analytic functions. In a region in space which is free of any currents and magnetized materials, the magnetic field fulfils the following two equations

$$\vec{\nabla} \cdot \vec{B} = 0, \ \vec{\nabla} \times \vec{B} = 0 \ . \tag{4.1}$$

This implies that we can express $\vec{B}$ either as the curl of the familiar vector potential $\vec{A}$ or as the gradient of a scalar magnetic potential V

$$\vec{B} = \vec{\nabla} \times \vec{A}, \quad \vec{B} = -\vec{\nabla}V . \tag{4.2}$$

For our two-dimensional problem the vector potential has only a z component. The x and y components of the magnetic field vector can be computed in two alternative ways

$$B_x = -\frac{\partial V}{\partial x} = \frac{\partial A_z}{\partial y}, \quad B_y = -\frac{\partial V}{\partial y} = -\frac{\partial A_z}{\partial x} . \tag{4.3}$$

If we define a complex potential function

$$\tilde{A}(x, y) = A_z(x, y) + iV(x, y) \tag{4.4}$$

then it is easy to prove that $\tilde{A}$ is an analytic function of the complex variable $\zeta = x + iy$ since the equations (4.3) are identical with the Cauchy-Riemann conditions which the real and imaginary part of an analytic function have to obey. So one can expand $\tilde{A}$ in a power series about the origin

$$\tilde{A}(x, y) = \sum_{n=0}^{\infty} c_n (x + iy)^n \tag{4.5}$$

which converges in the largest circle which contains neither current nor magnetized material. Taking the derivatives of (4.5), we obtain the multipole expansions for the components of the magnetic field vector[1].

Having shown that any two-dimensional field in vacuum can be expanded in a multipole series, we now compute this series explicitly for a very simple case, namely a single current-carrying wire. For this application it is convenient to use cylindrical instead of Cartesian coordinates. The magnet axis is chosen as the z direction of a cylindrical coordinate system (r, θ, z), shown in Fig. 4.2a. In the almost straight section of the magnet all current conductors are parallel to the z axis and can be considered as infinitely long since the transverse dimensions are small. Consider first a line current in the positive z direction which flows exactly on the z axis. The magnetic field lines are concentric circles around the z axis, so the field is purely azimuthal and has the familiar form

$$B_\theta = \frac{\mu_0 I}{2\pi r} .$$

---

[1]Another frequently used possibility is to define a complex magnetic field by the equation $\tilde{B} = B_y + iB_x$. Then the equations (4.1) are identical with the Cauchy-Riemann conditions and $\tilde{B}$ is seen to be analytic and can hence be expanded in a power series. We prefer the vector potential since a single scalar quantity, A_z in this case, is sufficient to compute the magnetic field pattern. For current-dominated magnets the vector potential is in fact very practical because the vector $\vec{A}$ is parallel to the current density $\vec{J}$. In the present application with all currents parallel or antiparallel to the z axis, the vector $\vec{A}$ has therefore just a z component. For conventional magnets the scalar potential is more adequate because the iron pole shoes are equipotential surfaces for $\mu \gg 1$.

The vector potential generated by the current has only a z component

$$A_z(r, \theta) = -\frac{\mu_0 I}{2\pi} \ln\left(\frac{r}{a}\right) .$$

The quantity a is an arbitrary length introduced here to make the argument of the logarithm dimensionless. It cancels in taking the radial derivative

$$B_\theta = -\frac{\partial A_z}{\partial r} = \frac{\mu_0 I}{2\pi r} .$$

If we more generally assume that the current is still parallel to the z axis but located at some point ($r = a, \theta = \phi$) in the r, θ plane, see Fig. 4.2b, the vector potential[2] has almost the same form as in the previous case:

$$A_z(r, \theta) = -\frac{\mu_0 I}{2\pi} \ln\left(\frac{R}{a}\right) . \tag{4.6}$$

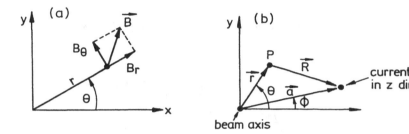

Figure 4.2: (a) Coordinate system for the multipole expansion. (b) Field calculation for a line current.

The only difference is that one has to replace r by the distance

$$R = \sqrt{a^2 + r^2 - 2ar\cos(\theta - \phi)}$$

between the location of the current and the general point $P = (r, \theta)$ at which we want to evaluate A_z. Let us first consider the case $r < a$. Then it is useful to write

$$R^2 = a^2[1 - (r/a)\exp(i(\theta - \phi))][1 - (r/a)\exp(-i(\theta - \phi))]$$
$$\ln(R/a) = \frac{1}{2}\ln[1 - (r/a)\exp(i(\theta - \phi))] + \frac{1}{2}\ln[1 - (r/a)\exp(-i(\theta - \phi))] .$$

Now we use the Taylor expansion of the logarithm

$$\ln(1 - \xi) = -\xi - \frac{1}{2}\xi^2 - \frac{1}{3}\xi^3 - \ldots - \frac{1}{n}\xi^n - \ldots$$

[2]In this section we follow partly the book *Superconducting Magnet Systems* by H. Brechna (1973) where the vector potential method is extensively used for calculating the field pattern.

which converges for arbitrary complex numbers ξ with $|\xi| < 1$. The vector potential and the field components are therefore for $r < a$

$$A_z(r, \theta) \;=\; \frac{\mu_0 I}{2\pi} \sum_{n=1}^{\infty} \frac{1}{n} \left(\frac{r}{a}\right)^n \cos[n(\theta - \phi)] \tag{4.7}$$

$$B_\theta(r, \theta) \;=\; -\frac{\partial A_z}{\partial r} = -\frac{\mu_0 I}{2\pi a} \sum_{n=1}^{\infty} \left(\frac{r}{a}\right)^{n-1} \cos[n(\theta - \phi)]$$

$$B_r(r, \theta) \;=\; \frac{1}{r}\frac{\partial A_z}{\partial \theta} = -\frac{\mu_0 I}{2\pi a} \sum_{n=1}^{\infty} \left(\frac{r}{a}\right)^{n-1} \sin[n(\theta - \phi)]$$

$$B_z(r, \theta) \;=\; 0 \;. \tag{4.8}$$

For certain applications, for instance in the computation of the local field inside the coil, one needs also the field components at distances $r > a$ from the axis. For this purpose we write R^2 in the form

$$R^2 = r^2 \left[1 - (a/r)\exp\left(i(\theta - \phi)\right)\right] \cdot \left[1 - (a/r)\exp\left(-i(\theta - \phi)\right)\right]$$

and obtain for $r > a$

$$A_z(r, \theta) \;=\; -\frac{\mu_0 I}{2\pi} \ln\left(\frac{r}{a}\right) + \frac{\mu_0 I}{2\pi} \sum_{n=1}^{\infty} \frac{1}{n} \left(\frac{a}{r}\right)^n \cos[n(\theta - \phi)] \tag{4.9}$$

$$B_\theta(r, \theta) \;=\; \frac{\mu_0 I}{2\pi r} + \frac{\mu_0 I}{2\pi a} \sum_{n=1}^{\infty} \left(\frac{a}{r}\right)^{n+1} \cos[n(\theta - \phi)]$$

$$B_r(r, \theta) \;=\; -\frac{\mu_0 I}{2\pi a} \sum_{n=1}^{\infty} \left(\frac{a}{r}\right)^{n+1} \sin[n(\theta - \phi)]$$

$$B_z(r, \theta) \;=\; 0 \;. \tag{4.10}$$

Note that the expressions (4.7) and (4.9) approach the same value in the limit $r \to a$, so $A_z(r, \theta)$ is well-defined for all r.

4.2   Generation of pure multipole fields

A single line current produces multipole fields of any order n which are of course of no practical use. To find out how one can generate a useful field we consider an arrangement of current conductors, running parallel to the z direction, which are mounted on a cylinder of radius a. A pure multipole field, containing just the single order $n = m$, is obtained inside the cylinder if the current distribution as a function of the azimuthal angle ϕ is given by

$$I(\phi) = I_0 \cos(m\phi) \;. \tag{4.11}$$

The statement is easily proved by computing the vector potential resulting from the current distribution (4.11):

$$A_z(r,\theta) = \frac{\mu_0 I_0}{2\pi} \sum_{n=1}^{\infty} \frac{1}{n} \left(\frac{r}{a}\right)^n \int_0^{2\pi} \cos(m\phi)\cos[n(\theta-\phi)]d\phi \ .$$

Using

$$\cos[n(\theta-\phi)] = \cos(n\theta)\cos(n\phi) + \sin(n\theta)\sin(n\phi)$$

and the orthogonality of the trigonometric functions one immediately sees that the integral vanishes unless $n = m$, so only a single term in the sum remains:

$$
\begin{aligned}
A_z(r,\theta) &= \frac{\mu_0 I_0}{2} \cdot \frac{1}{m} \left(\frac{r}{a}\right)^m \cos(m\theta) \\
B_\theta(r,\theta) &= -\frac{\mu_0 I_0}{2a} \left(\frac{r}{a}\right)^{m-1} \cos(m\theta) \\
B_r(r,\theta) &= -\frac{\mu_0 I_0}{2a} \left(\frac{r}{a}\right)^{m-1} \sin(m\theta) \ .
\end{aligned}
\tag{4.12}
$$

For $m = 1, 2, 3$ we obtain dipole, quadrupole and sextupole fields, respectively. These are shown in Fig. 4.3, together with the iron pole shoes of the corresponding normal magnets. The fields (4.12) are the so-called *normal multipole* fields. If we rotate the current distribution (4.11) by an angle of $\pi/(2m)$, we obtain a $\sin(m\phi)$ distribution leading to *skew multipole* fields. A skew dipole, for instance, has a horizontal field. Such magnets are positioned close to the vertically focusing quadrupoles to correct the particle orbit in the vertical plane. All other skew multipoles are quite undesirable in a circular accelerator. Skew quadrupole fields arise from an angular misalignment of the normal quadrupoles. They have the unpleasant feature of coupling horizontal and vertical betatron oscillations. A few correction quadrupoles, rotated by 45° around their axis, are usually needed to eliminate the coupling.

4.3 Approximation of pure multipole coils by current shells

Current distributions with a $\cos(m\phi)$ dependence are difficult to fabricate with a superconducting cable of constant cross section. In this section we discuss how they can be approximated with sufficient accuracy by current shells but other configurations like current blocks are also possible (Fig. 4.4). The quality of such an approximation can be judged from the general multipole expansion:

$$
\begin{aligned}
B_\theta(r,\theta) &= B_{ref} \sum_{n=1}^{\infty} \left(\frac{r}{r_0}\right)^{n-1} [b_n \cos(n\theta) + a_n \sin(n\theta)] \\
B_r(r,\theta) &= B_{ref} \sum_{n=1}^{\infty} \left(\frac{r}{r_0}\right)^{n-1} [-a_n \cos(n\theta) + b_n \sin(n\theta)] \ .
\end{aligned}
\tag{4.13}
$$

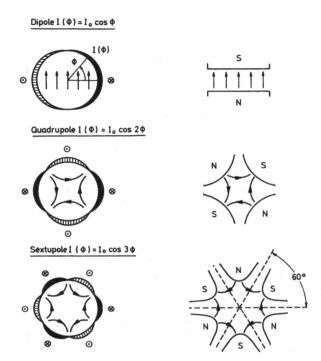

Figure 4.3: Generation of pure dipole, quadrupole and sextupole fields by $\cos(m\phi)$ current distributions and by conventional magnets with iron pole shoes.

Here r_0 is a *reference radius*, which should be in the same order as the maximum deviation of the protons from the centre axis of the magnet. A reasonable choice is about 2/3 of the inner-bore radius of the coil. The quantity B_{ref} is a reference field, for instance the magnitude of the main field at the reference radius, so $B_{ref} = B_1$ for a dipole and $B_{ref} = B_2(r_0) = g \cdot r_0$ for a quadrupole (g is the gradient). The b_n are called the *normal* multipole coefficients, the a_n are the *skew* coefficients. Alternatively, the complex magnetic field can be expanded in the multipole series

$$B_y + iB_x = B_{ref} \sum_{n=1}^{\infty} (b_n + i\,a_n) \left(\frac{x+iy}{r_0}\right)^{n-1} \qquad (4.14)$$

which permits easy computation of the magnetic field components in Cartesian coordinates. The expansions (4.13) and (4.14) are fully equivalent[3]. With the above

---

[3]In the American literature it is convention to start the multipole series with $n = 0$: $B_y + iB_x = B_{ref} \sum_{n=0}^{\infty} (b_n + ia_n)\,(x+iy)^n/r_0^n$. The multipole indices are therefore lower by one unit, so a normal dipole, quadrupole, sextupole is denoted by b_0, b_1, b_2, respectively, and similarly for the

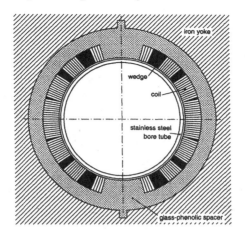

Figure 4.4: Approximation of a $\cos\phi$ distribution with current blocks in the RHIC dipole (Greene et al. 1995).

choice of B_{ref} the main coefficient is normalized to unity: $b_1 = 1$ in a dipole, $b_2 = 1$ in a quadrupole. The remaining coefficients should be very small for a good magnet, typically $|a_n|, |b_n| < 1 \cdot 10^{-4}$.

We observe that the ideal multipole coils of Fig. 4.3 have well defined symmetries. In a dipole coil, for any line current $+I$ at an angle ϕ, there exist three more currents: $+I$ at $-\phi$ and $-I$ at $\pi - \phi$ and $\pi + \phi$ (see Fig. 4.5a). The vector potential of these four currents, using Eq. (4.7), is

$$A_z(r,\theta) = \frac{2\mu_0 I}{\pi} \sum_{n=1,3,5,\ldots} \frac{1}{n} \left(\frac{r}{a}\right)^n \cos(n\phi)\cos(n\theta) . \qquad (4.15)$$

We can draw an important conclusion: a coil with dipole symmetry possesses only *normal multipoles* and no skew multipoles and only *odd* values of n appear.

Similarly, a coil with quadrupole symmetry (Fig. 4.6a) has again only normal multipoles and the 'allowed' orders are odd multiples of the lowest order 2: $n = 2, 6, 10, 14, \ldots$.

The simplest current shell arrangement with dipole symmetry is shown in Fig. 4.5b. We assume a constant current density J and compute the vector potential

skew multipoles. Moreover one has to watch out for the definition of the reference radius of the expansion. In the Tevatron and HERA magnets $r_0 = 1$ inch resp. $r_0 = 25$ mm was used, which is larger than the maximum beam deviation from the axis of the magnet, while at the SSC Laboratory a much smaller value of 10 mm was taken. With such a choice higher-order multipoles appear deceivingly small.

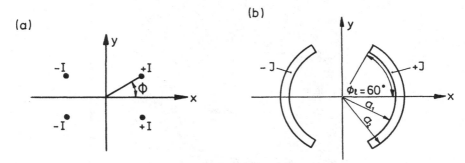

Figure 4.5:   (a) Four line currents with dipole symmetry.   (b) Simplest current-shell
arrangement for a dipole coil.

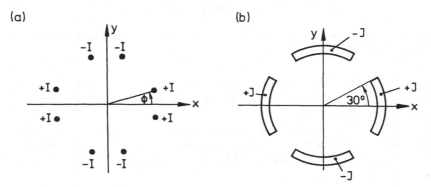

Figure 4.6:   (a) Line current arrangement with quadrupole symmetry.   (b) Simplest
current-shell arrangement for a quadrupole coil.

inside the coil using Eq. (4.15)

$$A_z(r,\theta) = \frac{2\mu_0 J}{\pi} \sum_{n=1,3,5,...} \frac{1}{n} \int_{a_1}^{a_2} \left(\frac{r}{a}\right)^n a\, da \int_0^{\phi_l} \cos(n\phi)\, d\phi \, \cos(n\theta) \quad .$$

Here ϕ_l is the limiting angle of the current shell and a_1, a_2 are its radii. The integrations can be done analytically[4].

$$A_z(r,\theta) = \frac{2\mu_0 J}{\pi} \sum_{n=1,3,5,...} \frac{r^n}{n^2(n-2)} (a_1^{2-n} - a_2^{2-n}) \sin(n\phi_l) \cos(n\theta) . \qquad (4.16)$$

---

[4]In a quadrupole made from current shells the sum in Eq. (4.16) extends over $n = 2, 6, 10, \ldots$.
For the lowest term $n = 2$ one has to integrate $1/a$ so $(a_1^{2-n} - a_2^{2-n})/(n-2)$ must be replaced by
$\ln(a_2/a_1)$.

For a thin current shell with $\Delta a = a_2 - a_1 \ll a = \frac{1}{2}(a_1 + a_2)$ the expression reduces to

$$A_z(r, \theta) = \frac{2\mu_0 J}{\pi} a \Delta a \sum_{n=1,3,5,\ldots} \frac{1}{n^2} \left(\frac{r}{a}\right)^n \sin(n\phi_l) \cos(n\theta) . \qquad (4.17)$$

The magnitude of the field of multipole order n is

$$B_n = \frac{2\mu_0 J}{\pi} \Delta a \frac{1}{n} \left(\frac{r}{a}\right)^{n-1} |\sin(n\phi_l)| . \qquad (4.18)$$

Choosing a limiting angle of $\phi_l = 60°$ the sextupole term $n = 3$ vanishes. Then the first non-vanishing higher multipole is the decapole $n = 5$. For typical coil dimensions the ratio

$$\frac{B_5}{B_1} = \frac{1}{5} \left(\frac{r}{a}\right)^4 \frac{|\sin 300°|}{\sin 60°}$$

is a few percent, two orders of magnitude larger than is tolerable. A single-layer current shell arrangement with constant current density is therefore too rough an approximation for a dipole coil. With two current shells, the sextupole and decapole can both be made to vanish by choosing a limiting angle of about 72° in the inner and of 36° in the outer layer. The Fermilab dipoles are built this way. There remain higher coefficients (b_7, b_9) which are in the order of 10^{-3}. A further reduction of these and all higher multipoles below the 10^{-4} level is possible by introducing longitudinal wedges into the inner and outer coil layer. The HERA, SSC and LHC magnets are constructed in this manner (Fig. 4.7).

A single-layer quadrupole coil (Fig. 4.6b) has a vanishing 12-pole ($n = 6$) but a b_{10} of about 2%. In the Fermilab and HERA quadrupoles two shells with additional wedges are used (Fig. 4.7b), and then most of the higher multipoles are in the 10^{-4} range or lower.

4.4 Influence of the iron yoke

The dipole and quadrupole magnets of an accelerator like the Tevatron or HERA are equipped with an iron yoke with a cylindrical inner bore which confines the magnetic field. Its influence on the field at the proton beam can be analyzed with the method of image currents provided the iron is not saturated and the permeability μ is uniform. Consider a current I at a radius a inside a hollow iron yoke whose inner surface is a cylinder of radius R_y. The effect of the iron on the inner field is equivalent to that of an image current I' which is located at the image radius $a' = R_y^2/a$:

$$I' = \frac{\mu - 1}{\mu + 1} \cdot I , \qquad a' = \frac{R_y^2}{a} . \qquad (4.19)$$

The image current I' runs parallel to the real current I and enhances the inner field. Fig. 4.8 shows the images of a single line current and of a current shell. In the latter

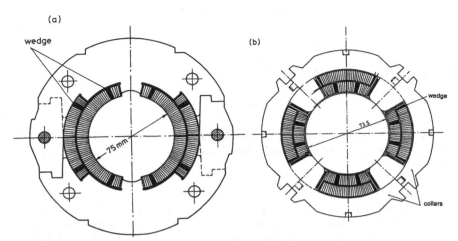

Figure 4.7:   (a) A two-shell dipole coil with longitudinal wedges in the inner and outer layer for improved field homogeneity. The coil is confined by non-magnetic collars. Coils of this type are used in the HERA, LHC and SSC magnets.   (b) Cross section of HERA quadrupole coil (Auzolle et al. 1986, 1989).

case the image current density is lower due to the increased area.

$$J' = \frac{\mu - 1}{\mu + 1} \cdot J \cdot \left(\frac{a}{R_y}\right)^4 \qquad \text{with} \qquad a = \sqrt{a_1 a_2} \ . \tag{4.20}$$

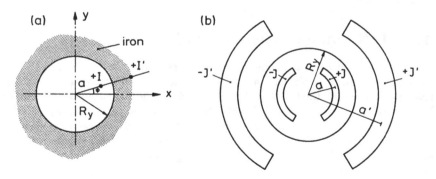

Figure 4.8:   (a) Image of a line current inside a hollow iron yoke (b) Image of a single-shell dipole coil.

For a single-layer dipole coil with concentric iron yoke the nth-order multipole field is

$$B_n(r,\theta) = \frac{2\mu_0}{\pi} \sin(n\phi_l) \frac{1}{n} \left[J\Delta a \left(\frac{r}{a}\right)^{n-1} + J'\Delta a' \left(\frac{r}{a'}\right)^{n-1} \right] .$$

Here the first term in the bracket is the coil contribution and the second term the iron contribution. Now

$$J'\Delta a' = \frac{\mu-1}{\mu+1} \cdot J\Delta a \cdot \left(\frac{a}{R_y}\right)^2 .$$

So for $n = 1$ and $\mu \gg 1$

$$(B_1)_{iron}/(B_1)_{coil} = (a/R_y)^2 . \tag{4.21}$$

As a simple example we consider just the inner coil shell in the HERA dipole, whose average radius is $a = 42.5$ mm. The iron yoke radius is $R_y = 88.4$ mm. In this case the relative iron contribution to the total dipole field on the axis is 19%.

For higher multipole orders n the iron contribution is much smaller:

$$(B_n)_{iron}/(B_n)_{coil} = (a^2/R_y^2)^n . \tag{4.22}$$

For the sextupole field B_3 this amounts to about 1.3% in the above example. The normalized sextupole coefficient $b_3 = B_3/B_1$ is reduced by about 18% because of the 19% iron contribution to the dipole field. In a two-layer coil, however, the sextupole and the allowed higher poles are modified by the yoke because the mirror image inverts the inner and outer coils. The limiting angles of the coil shells are adjusted in such a way that the sextupole vanishes when the coil is mounted inside the yoke. The collared coil without yoke has then a non-vanishing sextupole ($b_3 = 13 \cdot 10^{-4}$ in the HERA dipole coil). An important observation is that an unsaturated iron yoke does not create any new multipoles.

4.5 Saturation of iron yoke

The image current method fails when the yoke saturates since the permeability μ depends on position. Finite element programs are needed to compute the field pattern. With iron saturation the dependence of dipole field B_1 on current I is no longer linear and current-dependent sextupole and decapole coefficients arise. The saturation effects depend strongly on the proximity between coil and yoke. Three typical cases shall be considered.

4.5.1 'Warm-iron' dipole

In the Tevatron magnets (Fig. 4.9) the yoke is outside the cryostat and thus fairly far away from the coil. In this type of magnet saturation is almost negligible up to the critical current of the conductor. The iron contribution to the dipole field is about 10%; the field depends linearly on the current and no higher multipoles are observed.

4.5.2 'Cold-iron' dipole

For the Colliding Beam Accelerator project CBA and later the RHIC project, the Brookhaven laboratory has developed a dipole type whose coil is surrounded by a soft-iron yoke that is contained in the liquid helium cryostat (see Fig. 4.4). The yoke contributes about 35% to the central field, so a substantial saving in superconductor is possible. In the first version of the RHIC dipole strongly current-dependent sextupole and decapole coefficients were present but recently considerable progress has been achieved by increasing the thickness of the glass-phenolic spacer between coil and yoke and by optimizing the hole pattern in the yoke laminations. In the range of operation the saturation-induced multipoles deviate from the average by only $\pm 2.5 \cdot 10^{-4}$ for b_3 and $\pm 0.4 \cdot 10^{-4}$ for b_5, see Fig. 4.10.

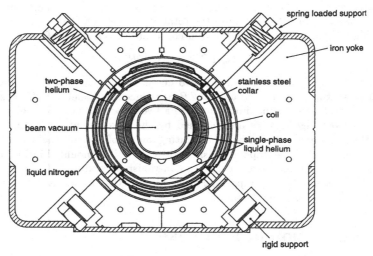

Figure 4.9: The Tevatron 'warm-iron' dipole (Tollestrup 1979).

4.5.3 'HERA-type' dipole

A third type, devised at DESY (Balewski, Kaiser, Schmüser 1984), combines the coil of the warm-iron design, clamped by non-magnetic collars, with an iron yoke inside the cryostat (Fig. 4.11). Here, the non-linearity in field versus current is quite moderate and the sextupole variation stays below $1 \cdot 10^{-4}$ for fields up to 6 T. The *transfer function* $B(I)/I$ of the HERA dipoles, normalized to its value at $I_0 = 2000$ A, is plotted in Fig. 4.12 against the coil current. The corresponding quadrupole curve is shown for comparison. The quadrupoles saturate earlier because their iron yoke is very slim. This design principle was adopted for SSC, UNK and LHC.

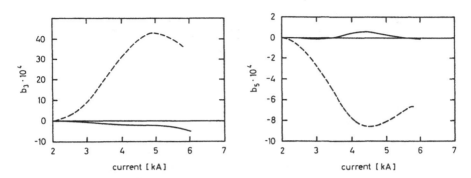

Figure 4.10: The current dependence of b_3 and b_5 in the first design (dotted curves) and in the final design (solid curves) of the RHIC dipole. The persistent-current contributions have been subtracted (R. Gupta, private communication).

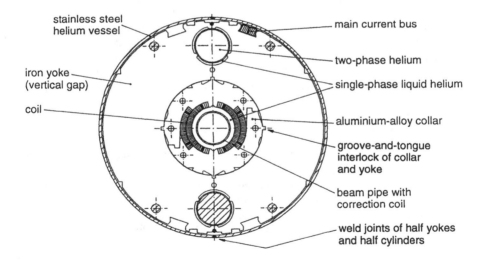

Figure 4.11: Cross section of the cryogenic part of the superconducting HERA dipole magnet (Wolff 1985, Schmüser 1985). The coil is clamped by an aluminium-alloy collar and then surrounded by a cold-iron yoke.

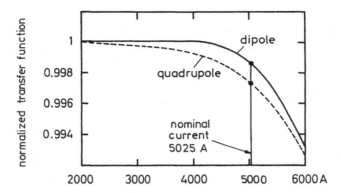

Figure 4.12: Current dependence of the normalized transfer functions $f_{norm}(I) = f(/I)/f(I_0)$ where $f(I)$ is defined by the relation $f(I) = B(I)/I$. For normalization an intermediate current $I_0 = 2000$ A was chosen at which persistent-current and yoke saturation effects are very small. The nominal operating current is 5025 A, corresponding to a dipole field of 4.68 T, a quadrupole gradient of 92 T/m and a proton energy of 820 GeV.

In the Large Hadron Collider the two counterrotating proton beams are bent and focused by twin-aperture magnets, having two coils of opposite polarity in a common iron yoke. A cross section is shown in Fig. 4.13. At the design field of 8.36 T, corresponding to a proton energy of 7000 GeV, there is significant iron saturation in the centre region. The resulting normal quadrupole component is minimized by a suitable hole pattern in the iron yoke laminations and by ferromagnetic inserts in the collars. The remaining b_2 of about 2 units of 10^{-4} at high field is compensated by the quadrupole magnets in the LHC ring which are fed by independent power supplies.

4.6   End field

Due to the complexity of the current distribution the field calculation at the coil end is far more complicated than in the straight section. No analytic formulation exists. Concerning the overall field quality, the end fields play a minor role for long magnets. Nevertheless the perturbations may be strong enough to require a compensation in the straight section. For a simple coil head configuration as shown in Fig. 4.14a, a large negative sextupole field is obtained (Fig. 4.14b). In the Tevatron dipoles this end-field sextupole is compensated by a purposely introduced positive sextupole in the straight section. Another disadvantage of the simple configuration is the fact that the field at the conductors in the return region is enhanced by about 10% above the central field. This is particularly worrysome since the windings in the coil head cannot be clamped as tightly as in the straight section and are thus more vulnerable to Lorentz-force induced motion with the danger of quenching. In more recent designs

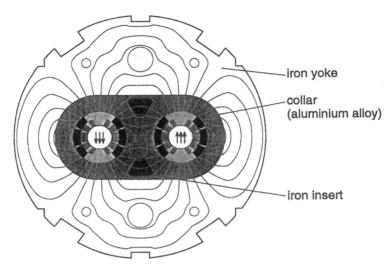

Figure 4.13: A cross section of the twin-aperture LHC dipole with computed field lines in the iron. One can see very clearly that the field pattern is strongly influenced by the arrangement and size of the holes in the iron yoke and also by the iron-inserts in the collars. For the RHIC and LHC magnets great care was taken to optimize the hole pattern in the yoke for minimum field distortions. This cannot be done analytically but needs numerical optimization codes. (Courtesy R. Perin).

the windings in the coil head are spread out by epoxy-fibreglass spacers. This is shown schematically in Fig. 4.15a.

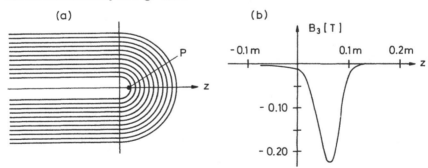

Figure 4.14: (a) Unwrapped view of simple coil head. The highest field in the coil is at the point P. (b) Sextupole field in the simple coil head configuration (dipole field is 5 T).

With a suitable choice of spacers the sextupole produced by the coil ends has both positive and negative values (Fig. 4.15b) and averages to zero. The basic principle is easy to understand. The windings at azimuthal angles below 30° produce a positive sextupole while those above 30° yield a negative sextupole. If a single-layer coil spans the angular range from 0 to 60° in its straight section, the two contributions cancel. By separating the return sections for the two groups of conductors one can accomplish a bipolar sextupole in the coil head region.

An additional benefit of the spacers in the head region is a reduction in the local field enhancement. If moreover the iron yoke does not cover the return region but terminates at the end of the straight section, the highest field point moves away from the coil head into the straight section where the conductors are firmly confined by the collars. This is an important step towards achieving a good quench performance.

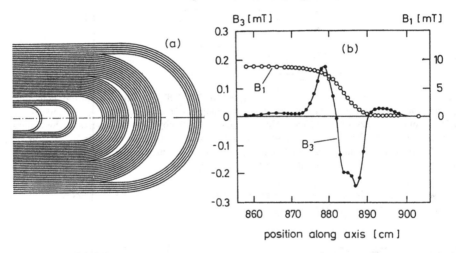

Figure 4.15: (a) Schematic view of coil head with spacers. (b) Measured dipole and sextupole end field of a HERA dipole for a current of 10 A in the normal state.

A knowledge of the end-field is highly desirable for both improving the overall field quality and reducing the local field in the coil. A straightforward but cumbersome method for calculating end fields is by means of the Biot-Savart law. The field $d\vec{B}$ at a point P which is produced by a conductor element $d\vec{l}$ carrying the current I is given by

$$dB = \frac{\mu_0 I}{4\pi} \cdot \frac{d\vec{l} \times \vec{r}}{r^3}$$

where $\vec{r}$ is the vector from P to the center of $d\vec{l}$. For a straight conductor parallel to the z axis that is infinitely long in the negative z direction and ends at $z = 0$ the magnetic field $\vec{B}$ at the point P is obtained by integration

$$\vec{B} = \frac{\mu_0 I}{4\pi R}\left(1 + \cos\theta_0\right)\vec{e}_z \times \frac{\vec{R}}{|\vec{R}|} .$$

Here $\vec{e}_z$ is a unit vector in z direction, $\vec{R}$ the position vector of P and θ_0 the angle between $\vec{R}$ and the z axis. In the coil head, the curved parts have to be approximated by a sequence of short straight sections.

There is a more elegant method, basically a piecewise multipole expansion. It rests on the assumption that the path of each conductor can be parametrized by a simple curve in space. A reasonable design is obtained by placing the conductors on a cylinder with constant radius a in such a way that the coil ends are half circles when unwrapped from the cylinder. Let the origin of the circle be at $z = z_0$. Then the coordinates of the k-th conductor can be parametrized in terms of an azimuthal angle $\phi_k(z)$ that depends on the longitudinal coordinate z.

$$\phi_k(z) = \begin{cases} \phi_k(0) & \text{for } z \leq z_0 \\ \dfrac{\pi}{2} - \sqrt{\left(\dfrac{\pi}{2} - \phi_k(0)\right)^2 - \left(\dfrac{z - z_0}{a}\right)^2} & \text{for } z_0 \leq z \leq z_{max} . \end{cases} \quad (4.23)$$

Here $z_{max} = z_0 + \left(\frac{\pi}{2} - \phi_k(0)\right)a$. For the field integral only the longitudinal component of the current I has to be considered since the azimuthal component produces a field (B_x, B_y) that is antisymmetric in z and cancels in the z integration.

$$I_z(k, z) = \begin{cases} I & \text{for } z \leq z_0 \\ I \cdot \dfrac{\phi_k(z) - \pi/2}{\phi_k(0) - \pi/2} & \text{for } z_0 \leq z \leq z_{max} . \end{cases} \quad (4.24)$$

Assuming that the left-right and top-bottom symmetry of the current conductors is preserved in the coil head region, one obtains just the allowed normal multipole fields B_n with $n = 1, 3, 5, \ldots$.

$$B_n(z) = \frac{2\mu_0}{\pi a}\left(\frac{r_0}{a}\right)^{n-1}\sum_{k=1}^{N} I_z(k, z)\cos n\phi_k(z) . \quad (4.25)$$

Of particular interest is the integral over the dipole component since it defines the magnetic length

$$l_m = \frac{1}{B_0}\int_{-\infty}^{+\infty} B_1(z)\,dz . \quad (4.26)$$

Here B_0 is the value of the dipole field in the central region of the magnet. The contribution of the coil ends to the harmonic coefficients of the entire magnet is obtained by integrating Eq. (4.25) along the z axis and dividing by $B_0 \cdot l_m$. In real magnets the left-right and top-bottom symmetries are usually disturbed in the coil ends, for instance through the current leads and the internal coil connections. Therefore also unallowed multipoles appear.

While it may appear somewhat dubious from a mathematical point of view to apply a two-dimensional harmonic expansion to a genuine three-dimensional problem, this procedure is very useful for obtaining a first quick estimate and for optimizing a coil design. The method has been considerably extended by Ijspeert, Tortschanoff and Wolf (1991) for the computation of multipole corrector magnets. The z-integrated multipole fields are found to be quite accurate.

A genuine three-dimensional treatment requires the use of finite-element computer programs like OPERA, TOSCA or ANSYS[5] which take into account the iron yoke and its saturation. The main task with these programs is to correctly implement the coil structure of the magnet and to find an adequate mesh for the calculation. The computing times are usually rather long, so these programs are not so useful in the design phase. As an example, the computed end field sextupole and decapole of an SSC dipole are depicted in Fig. 4.16.

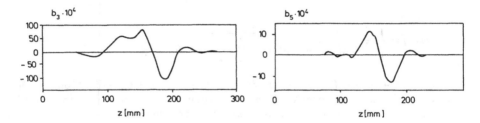

Figure 4.16: End field sextupole and decapole in an SSC dipole as computed with OPERA (Bliss et al 1993). (© 1993 IEEE)

S. Russenschuck (1995) has developed an elaborate optimization code called 'ROX-IE' which is used to optimize the conductor arrangement in the LHC dipoles, to design the coil ends and to provide the data for manufacturing of the coil-head spacers on a five-axis milling machine. An example will be presented in Fig. 10.3.

References

A. Auzolle et al., *Superconducting quadrupoles for HERA*, ICFA Workshop on Superconducting Magnets and Cryogenic Systems, Brookhaven 1986, BNL report 52006, p. 195, and: *First industry made superconducting quadrupoles for HERA*, IEEE Trans. **MAG-25** (1989) 1660

D.W. Bliss et al., *Magnetic and mechanical considerations in the design of the SSC collider dipole magnet end region*, IEEE Trans. **ASC-3** (1993) 692

---

[5]OPERA and TOSCA are registered trademarks of Vector Fields Ltd, ANSYS is a registered trademark of Swanson Analysis Systems, Inc.

H. Brechna, *Superconducting Magnet Systems*, Springer, Berlin 1973

A.F. Greene et al., *The magnet system for the Relativistic Heavy Ion Collider (RHIC)*, 14th Int. Conf. on Magn. Techn. MT-14, Tampere, Finland 1995

S. Russenschuck, *A computer program for the design of superconducting accelerator magnets*, Proc. ACES Symposium, Monterey, California 1995

A.V. Tollestrup in: ECFA study of an ep facility for Europe, DESY Report 79/48 (1979)

Chapter 5

Mechanical Accuracies and Magnetic Forces

5.1 Mechanical tolerances

In Chap. 4 we have shown that a good approximation of the ideal $\cos\phi$ current distribution of a dipole can be realized in two ways: (a) by using two or more coil layers with properly adjusted limiting angles; (b) by making the current distribution in a single-shell coil non-uniform with the help of longitudinal wedges which create current-free regions. Method (a) was used in the Tevatron magnets, method (b) in the CBA and RHIC magnets. For fields of 5 Tesla or more it is advantageous to combine both methods and construct coils with two shells plus additional wedges. The magnets for HERA, SSC, LHC are made this way. Although one is not completely free in the choice of the coil angles because of the fixed cable thickness, an excellent field homogeneity can still be achieved. An example for such an optimized structure is the HERA dipole coil (see Fig. 4.7) whose straight-section harmonics are listed in Table 5.1. It should be noted that in the SSC and LHC magnets cables of different

Table 5.1: Calculated multipole coefficients of the HERA dipole; the reference radius is $r_0 = 25$ mm. The numbers are given in units of 10^{-4}. Iron yoke saturation and persistent-current effects have been neglected.

b_3	b_5	b_7	b_9	b_{11}	b_{13}	b_{15}
0.0	0.9	0.3	-0.6	-0.1	-0.4	-0.3

thickness and width are used for the inner and outer coil layer while for HERA both layers were made from the same cable.

There is not much use to strive for even smaller harmonics than quoted in Table 5.1 since in real magnets the manufacturing tolerances readily outweigh these theoretical numbers. The dimensional accuracy is usually in the order of a few hundredths of a millimetre. For a single-layer dipole coil whose field is easily computed analytically we want to discuss a typical geometrical error and evaluate its influence on field quality. If the limiting angle ϕ_l of the current shell differs from $60°$ the sextupole coefficient is no longer zero. It can be computed from Eq. (4.18)

$$b_3 = \frac{1}{3} \left(\frac{r_0}{a} \right)^2 \frac{\sin(180° + 3\delta\phi_l)}{\sin(60° + \delta\phi_l)}$$

where $\delta\phi_l$ is the angular error. The condition that $|b_3| \leq 1 \cdot 10^{-4}$ requires $\delta\phi_l \leq 0.25$ mrad, i.e. the arc length of a half coil must be accurate to 0.01 mm for an average coil radius $a = 40$ mm.

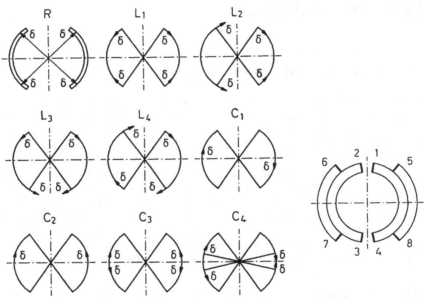

Figure 5.1: The transformations R, L and C leading to geometric distortions of a dipole coil and the eight limiting angles and shim positions in a two-shell dipole coil.

In the general case, various conceivable distortions of the coil geometry can be characterized by a set of transformations acting on an ideal coil[1]. They are depicted in Fig. 5.1. The radial transformation R increases the inner and outer radius of the current shell by the same amount $\delta R = 0.1$ mm. Since the dipole symmetry is preserved, no unallowed multipoles arise but the allowed poles $b_1, b_3, b_5, \cdots$ are changed. The transformations labeled L modify the limiting angles. L_1 preserves both the top-bottom and the left-right symmetry. Consequently, no new multipoles arise

---

[1]These transformations, which influence either only the normal multipoles b_n or the skew poles a_n, were introduced by A.V. Tollestrup for the analysis of the Tevatron magnets.

but there are changes δb_1, δb_3, δb_5, $\cdots$ in the allowed multipoles. The transformation L_2 (L_3) leads to a left-right (top-bottom) asymmetry and creates non-vanishing δb_n (δa_n). Transformation L_4 may appear like a rotation of the whole coil but in fact the mid-plane stays invariant. Here $\delta a_n \neq 0$. The transformations labeled C affect the centre plane. C_1, C_2 rotate resp. shift the mid-plane between the top and bottom half-coils while C_3, C_4 introduce a symmetric or asymmetric gap in the mid-plane.

Table 5.2: Effect of geometrical distortions on the harmonic coefficients of the HERA dipole coil. The effect has been calculated for the transformations R, L and C, sketched in Fig. 5.1. The magnitude of the radial or azimuthal deformation is always δl =0.1 mm. The harmonics refer to a radius r_0 = 25 mm and are quoted in units of 10^{-4}. The numbers in brackets are for a coil without iron yoke.

transformation	n	inner coil δa_n	inner coil δb_n	outer coil δa_n	outer coil δb_n
R	1		-9.0 (-14.6)		-3.5 (-7.1)
	3		2.2 (2.0)		-1.7 (-2.2)
L_1	1		6.4 (6.8)		1.7 (1.6)
	3		2.4 (3.1)		1.7 (2.2)
	5		-1.1 (-1.4)		0.3 (0.4)
	7		0.3 (0.4)		- (-)
L_2	2		7.5 (9.1)		2.4 (2.7)
	4		-1.2 (-1.5)		0.9 (1.2)
L_3	2	1.2 (1.4)		-1.8 (-2.1)	
	4	2.3 (2.9)		0.3 (0.3)	
	6	-0.5 (-0.6)		0.2 (0.3)	
L_4	1	-5.8 (-6.0)		-3.6 (-3.4)	
	3	4.3 (5.4)		-0.3 (-0.4)	
C_1	1	-8.1 (-8.4)		-4.0 (-3.7)	
	3	-2.0 (-2.6)		-1.5 (-1.8)	
	5	- (-)		-0.3 (-0.4)	
C_2	2	5.3 (6.5)		2.7 (3.1)	
	4	0.4 (0.4)		0.7 (0.9)	
C_3	1		-3.7 (-3.7)		-0.8 (-0.8)
	3		-4.9 (-6.2)		-1.2 (-1.3)
	5		-1.6 (-2.0)		-0.5 (-0.6)
C_4	2		5.7 (6.9)		1.3 (1.4)
	4		3.0 (3.8)		0.8 (1.0)
	6		0.8 (1.0)		0.3 (0.3)

The changes of the harmonic coefficients which result from these transformations can be readily computed by replacing each cable in the coil by a sufficient number of straight current conductors (24 were taken in the case of HERA) and using Eq. (4.8) to compute the field. Table 5.2 lists the harmonic changes in the HERA dipole coil due to the transformations of Fig. 5.1.

The above transformations are the basis of an error analysis which is applied in the correction of geometrical errors arising in the industrial production of the coils. This is done by *shimming*, that is by properly adjusting the thickness of the thin stainless-steel strips that are placed between the insulated coil package and the collar at the eight limiting angles (see Fig. 5.1). Any displacement at these positions can be characterized by a vector $\vec{\delta l}$. The skew and normal multipole coefficients a_n, b_n for $n = 2, 3, 4, 5$ turn out to be particularly sensitive to the resulting angular errors. The variation in harmonic coefficients is obtained from the matrix equation

$$
\vec{\delta h} \equiv
\begin{pmatrix}
\delta a_2 \\
\delta a_3 \\
\delta a_4 \\
\delta a_5 \\
\delta b_2 \\
\delta b_3 \\
\delta b_4 \\
\delta b_5
\end{pmatrix}
= M \cdot
\begin{pmatrix}
\delta l_1 \\
\delta l_2 \\
\delta l_3 \\
\delta l_4 \\
\delta l_5 \\
\delta l_6 \\
\delta l_7 \\
\delta l_8
\end{pmatrix}
\tag{5.1}
$$

The matrix M is related in a straightforward way to the transformations discussed above. For the HERA dipole coil without yoke it is given by

$$
M =
\begin{bmatrix}
0.35 & 0.35 & -0.35 & -0.35 & -0.525 & -0.525 & 0.525 & 0.525 \\
1.35 & -1.35 & 1.35 & -1.35 & -0.10 & 0.10 & -0.10 & 0.10 \\
0.725 & 0.725 & -0.725 & -0.725 & 0.075 & 0.075 & -0.075 & -0.075 \\
-0.025 & 0.025 & -0.025 & 0.025 & 0.10 & -0.10 & 0.10 & -0.10 \\
-2.275 & 2.275 & 2.275 & -2.275 & -0.675 & 0.675 & 0.675 & -0.675 \\
-0.775 & -0.775 & -0.775 & -0.775 & -0.55 & -0.55 & -0.55 & -0.55 \\
0.375 & -0.375 & -0.375 & 0.375 & -0.30 & 0.30 & 0.30 & -0.30 \\
0.35 & 0.35 & 0.35 & 0.35 & -0.10 & -0.10 & -0.10 & -0.10
\end{bmatrix}
$$

The measured harmonic coefficients of a collared coil are used to compute the required shims by applying the inverse matrix

$$
\vec{\delta l} = M^{-1} \cdot \vec{\delta h} .
$$

In the calculation of the matrix elements it has been assumed that a variation of coil arc length is distributed uniformly among all conductors. The thickness of the wedges is kept constant as they are made from solid copper.

Apart from the coil angles the position of the mid-plane between the top and bottom half coil of a magnet may move as a consequence of fabrication tolerances. This is actually quite likely due to the fact that there is no mechanical element in the coil clamping structure which keeps the mid-plane in place, so basically this plane is floating. During collaring the half-coils are compressed like two springs and, depending on the relative size of the upper and lower half-coil and on their elastic moduli, the mid-plane adjusts itself. A displacement of only 0.1 mm produces already a skew quadrupole (a_2) of 6.5 units of 10^{-4}, as can be seen from Table 5.2. Similarly, a tilt of the centre plane (transformation C_1) generates a skew dipole (a_1) of -8.4 units and a skew sextupole (a_3) of -2.6 units.

A method to reduce these distortions is to measure the arc length and the modulus of elasticity on each side of a half-coil at many positions along the length and to combine half-coils that match in arc length and elastic modulus. This tedious procedure was applied for the HERA magnets and for prototype SSC magnets. In spite of this effort, mid-plane shifts of up to 50 μm remained resulting in harmonic distortions of considerable size, especially skew quadrupoles a_2 and sextupoles a_3. A detailed analysis by Ogitsu et al. (1993) revealed a good correlation between the measured skew sextupoles in SSC dipoles and those calculated from the known coil dimensions. The correlation for the skew quadrupole was less pronounced. Systematically different skew sextupoles were found in SSC magnets from different production sites. The reason may have been asymmetries in the molds used for curing the coils.

From the above considerations we conclude that mechanical accuracies in the 20 μm range are needed to satisfy the field homogeneity requirements. Such narrow tolerances are difficult to accomplish by conventional machining, in particular for magnets of 9 or even 15 m length. Using precision-stamped laminations to assemble the tooling for coil winding and baking and also for the collars which clamp the finished coil one can achieve the required precision at any cross section of the coil.

Another type of distortion is an off-centring of the coil in the cylindrical yoke. The field of a dipole acquires in this case a quadrupole component. A horizontal displacement of 0.1 mm leads to a normal quadrupole coefficient of $(1 - 2) \cdot 10^{-4}$, for a vertical shift one gets a skew quadrupole of similar size. In a cold-iron magnet, a centring of better than 0.1 mm is no problem since the collars and the yoke can be interlocked by precisely stamped tongues and grooves (see Fig. 4.11) but in a warm-iron dipole many supports and a precise alignment are needed. An interesting aspect is the observation made at the SSC laboratory that an off-centre arrangement of the dipole magnet inside the steel vacuum vessel of the cryostat may lead to a skew quadrupole at high excitation when the iron yoke saturates and is insufficient to contain the entire magnetic flux.

5.2 Multipole measurements as a means of quality assurance

The quality assurance tests performed on all HERA dipoles and quadrupoles yielded a wealth of information on the accuracies which can be achieved in a large-scale industrial production of superconducting magnets. One of the most sensitive tests on

the precision of the coil cross section and of the placement of the current conductors
is provided by the multipole measurements. Fig. 5.2 shows the normal and skew
multipole coefficients of the HERA dipoles at a current of 5000 A, corresponding to
a field of 4.66 Tesla. Most of the coefficients are very small and well within the limits
of $\pm 0.5 \cdot 10^{-4}$, which were used in the particle tracking programs for determining
the dynamic aperture (maximum stable beam size, see also Sec. 9.1) of the HERA
storage ring. So the general conclusion from the data in Fig. 5.2 is that industry has
been able to conform with the stringent requirements on mechanical precision.

Two coefficients show a larger scattering: the normal sextupole b_3, which is particularly sensitive to slight changes in the limiting angles of the coil shells, and the
skew quadrupole a_2, which may arise from an up-down asymmetry of the coil. The
dipoles have been sorted in the accelerator in order to minimize the sextupole effects.
At the production plants the two half-coils of a dipole were matched with respect
to their elastic moduli in order to minimize top-bottom asymmetries. The collared
coils were measured at room temperature before installation in the iron yoke and
cryostat. If the sextupole turned out too large, the collars were opened, shims were
added and then the collars were closed again. Since cryogenic measurements are very

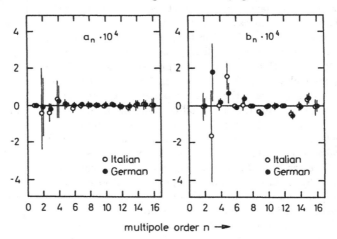

Figure 5.2: The normal (b_n) and skew (a_n) multipole coefficients of the HERA dipoles at
5000 A, corresponding to a field of 4.66 T and a proton energy of about 800 GeV. Plotted
are the average values with their rms standard deviations from 200 magnets of Italian and
German production. The data have been averaged over the whole length of the magnets,
including the coil heads. The reference radius is $r_0 = 25$ mm, 2/3 of the inner-bore radius
of the coil.

time consuming there is considerable interest whether measurements in the normal-
conducting coils at room temperature are sufficiently accurate for determining the
field quality. In Fig. 5.3 we show the correlation between the harmonic coefficients

measured in 'warm' dipole coils and those determined in complete magnets at 5000 A in the superconducting state[2]. Generally a good correlation is observed and from the spread of the data one can conclude that room-temperature measurements are adequate to ensure the proper geometry of the coils. This is very important for quality control during production since field measurements without the need of cooling can be performed by the coil manufacturer.

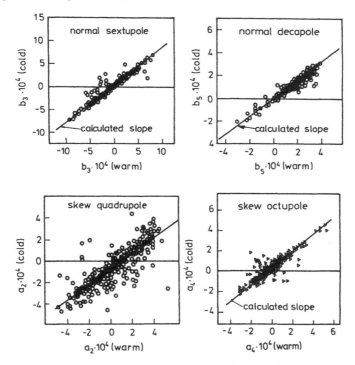

Figure 5.3: Correlation between room-temperature and cryogenic measurements of the multipole coefficients in the HERA dipoles. The 'warm' measurements were made in collared coils without iron yoke. For this reason the slope in the correlation plots is less than 1. Moreover, a systematic sextupole offset of $13 \cdot 10^{-4}$ in the 'warm' data (see Sect. 4.4) has been subtracted.

The magnets of the heavy ion collider RHIC are wound from a superconducting cable with narrow dimensional tolerances (see also Chap. 3). Moreover the coil consists of a single layer only. The geometrical distortions tend to be smaller than in the HERA magnets which were produced ten years earlier. The 'warm-cold' correlation

---

[2]At room temperature the current is limited to 10–20 A. To improve sensitivity an alternating current of 11 Hz was used and the induced signal was detected with a lock-in analyzer.

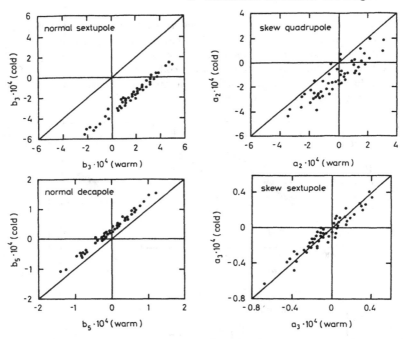

Figure 5.4: Correlation between room-temperature and cryogenic measurements of the multipole coefficients in the RHIC dipoles (R.C. Gupta, private communication). The systematic offset in the sextupole and decapole is caused by the fact that the cryogenic data were taken at high current where yoke saturation comes into play.

in these magnets is indeed rather good (Fig. 5.4) and allows to restrict cryogenic measurements to a 10% fraction of the magnets without sacrificing field quality.

The techniques for multipole measurements are described in Appendix A.

5.3   Field integral and field orientation

What counts for the deflection and focusing of a particle beam is the integral of dipole field and quadrupole gradient over the length of the magnet. An absolute determination of the integrated dipole field is possible by a longitudinal scan with a nuclear magnetic resonance (NMR) probe plus an additional Hall probe that covers the inhomogeneous end field (Preissner et al. 1990). For a quadrupole one can stretch a wire along the axis and move it perpendicular to the field by precision-tables, see Appendix B. The magnetic flux swept in the motion translates into a time integral of the induced voltage. Both methods permit accuracies in the 10^{-4} range. The field integral distribution of the HERA magnets is plotted in Fig. 5.5. The dipoles from two production lines (Ansaldo/Zanon in Italy and Brown Bovery in Germany) turned out systematically different by $1.9 \cdot 10^{-3}$ although both were built according to the same drawings. Half of this difference is caused by different magnetic lengths (8825.7 ± 1.7 mm vs. 8833.5 ± 2.0 mm), the other half by different central fields ($B/I = 0.9328 \pm 0.0005$ T/kA resp. 0.9336 ± 0.0005 T/kA). In the HERA machine, correction magnets are used to compensate the systematic offset between the two sets of dipoles.

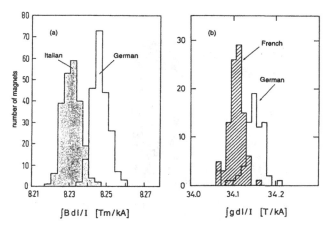

Figure 5.5: (a) Field integral of all HERA dipoles, normalized to coil current. (b) Integrated gradient of all quadrupoles, normalized to coil current (Brück et al. 1991).

In order to keep the unavoidable closed-orbit distortions in the accelerator ring within tolerable limits (typically 1 mm rms in a large machine), the field orientation of the dipoles with respect to the ring plane has to be kept within a fraction of a milliradian and the axes of the quadrupoles must be aligned with an accuracy of better than 0.25 mm with respect to the nominal orbit. Contrary to conventional magnets with iron pole shoes, the centre axis and the field direction of superconducting magnets are not directly observable with theodolites and leveling instruments when the cryostat is closed. A tedious procedure is needed to transfer the data from magnetic measurements to optical reference targets which are mounted on the outer cryostat vessel and which can later be utilized for surveying the magnets in the accelerator tunnel.

The dipole field direction can be determined with a device that combines two orthogonal Hall probes and a gravitational sensor with electronic readout. The precision is a fraction of a milliradian. The collared dipole coil by itself is not a stiff structure and may be bent or twisted along its length. The mechanical stability is provided by a stainless-steel cylinder welded around the iron yoke. The average twist of the HERA dipoles is 0.06 ± 0.2 mrad/m. An interesting method is applied at Brookhaven to determine the position of the axis of quadrupoles and sextupoles and the field orientation at any place along the magnet. The field pattern of these magnets is made visible by inserting a cell with a ferrofluidic colloidal solution and observation with polarized light (Trbojevic et al. 1995). An image is shown in Fig. 5.6. A similar procedure was used for the alignment of quadrupoles at SLAC, see Cobb and Muray (1967) who present also the mathematical theory of the method. The axis of the RHIC quadrupoles is measured locally with an rms precision of 0.05 mm. At HERA a stretched-wire system with comparable resolution was used for a global determination of the axis, see Appendix B.

Figure 5.6: Image of a cell with a ferrofluidic colloidal solution in a RHIC quadrupole, viewed with polarized light (courtesy D. Trbojevic).

5.4   Magnetic forces

As an example we consider a two-layer dipole coil of the HERA or Tevatron size which is excited to a field of 5 T. The radial and azimuthal components of the Lorentz forces acting on the superconducting cable are plotted in Fig. 5.7 as a function of conductor number. For simplicity the wedges inside the coil are neglected. In the horizontal plane the radial forces dominate. A conductor in the inner coil shell is pushed outwards with a force of 13000 N per metre length, a conductor of the outer shell is pulled inwards with 6000 N/m. Summing over all windings we obtain a horizontal force of 10^6 N for a 1-m-long coil section.

Close to the limiting angles of the coil shells, the azimuthal forces dominate. They are directed towards the median plane and have the tendency to compress the coil package and move it away from its end stop at the collars. Such a movement has to be inhibited to avoid frictional heating and quenching.

The azimuthal motion in a half coil will be studied using a simplified model proposed by Tollestrup (1979). The coil package is considered as a system of compressed springs which are confined between two end stops (Fig. 5.8). The azimuthal magnetic force acts between any two springs and is roughly proportional to the conductor

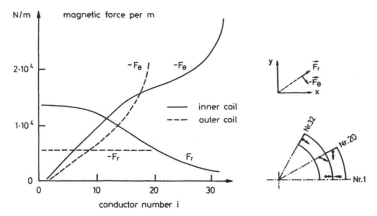

Figure 5.7: Azimuthal and radial magnetic forces per metre length at a field of about 5 T.

number i as can be seen from Fig. 5.7: $F_i = \alpha i$. Let k be the spring constant and δX_i the displacement of spring number i. Then we have the following system of difference equations:

$$k(\delta X_{i+1} - \delta X_i) - k(\delta X_i - \delta X_{i-1}) = -F_i = -\alpha i . \qquad (5.2)$$

This system can be solved by assuming for the displacements X_i a third-order polynomial

$$\delta X_i = a i^3 + b i^2 + c i + d . \qquad (5.3)$$

Substituting into (5.2) and using the boundary condition $\delta X_0 = \delta X_{N+1} = 0$ (we require that the package does not lift off at the end stops) one gets

$$a = -\frac{\alpha}{6k} , \qquad b = 0 , \qquad c = \frac{\alpha}{6k}(N+1)^2 , \qquad d = 0 .$$

The displacement of conductor i is therefore

$$\delta X_i = \frac{\alpha}{6k} i [(N+1)^2 - i^2] . \qquad (5.4)$$

The constant α is about 840 N for the inner shell. The spring constant k is related to the elastic modulus Y of the coil package. From the series measurements on the HERA dipole coils one obtained $Y = (27 \pm 4)$ GPa. This corresponds to a spring constant $k = 1.6 \cdot 10^{11}$ N/m for a 1-m coil section. The maximum displacement of the conductors under the influence of the Lorentz forces amounts to about 0.011 mm and is obtained for conductor 20.

The resulting magnetic force F_{end} which tries to pull the coil away from the end stop at the key angle is given by

$$F_{end} = k \delta X_N = \frac{\alpha}{6} N(2N+1) = 2.7 \cdot 10^5 \text{Nm}^{-1} . \qquad (5.5)$$

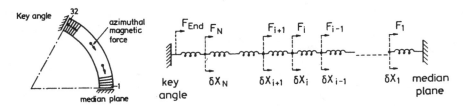

Figure 5.8: Precompressed coil section and equivalent spring model. (After Tollestrup 1979).

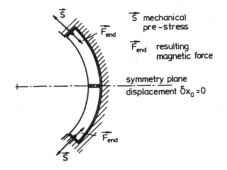

Figure 5.9: Simplified picture of precompressed coil. S: mechanical prestress, F_{end} : resulting magnetic force.

The coil is 10 mm wide, so this corresponds to a (negative) pressure of 27 MPa at the key angle. Choosing a mechanical prestress S well in excess of this value one can avoid a coil motion at the key angle (see Fig. 5.9). The mechanical pre-compression in the HERA magnets is equivalent to a (positive) pressure of about 60 MPa (600 bar). One should bear in mind that the calculation is oversimplified since any friction between the coil and clamp has been neglected and the coil package has been treated as an elastic spring which is far from reality.

The collars which confine the coils have to be strong enough to apply the required prestress on the coils and to take up the huge Lorentz forces. Figure 5.10 shows the computed deformation of the aluminium-collared HERA dipole coil in two states: at room temperature and zero field and in the superconducting state at 6 Tesla. The counter force of the strongly compressed coil deformes the circular-shaped collar into an upright ellipse if no magnetic forces are present. Part of this vertical lengthening is removed when the collared coil is installed in the iron yoke. With increasing excitation the magnetic forces start to dominate and generate the horizontal elliptical deformation depicted in Fig. 5.10b. Although the computed deformations exceed the limits given in Sect. 5.1 they cause only a small sextupole (less than $1 \cdot 10^{-4}$) since an

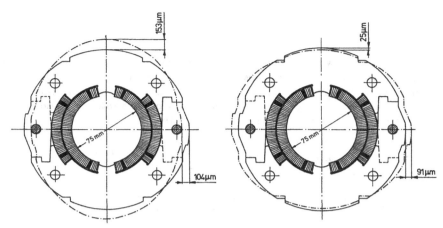

Figure 5.10: Calculated deformation of the collared HERA coil: (a) at room temperature without magnetic field and (b) in the superconducting state at 6 T (G. Meyer, private communication). The collar material is aluminium AlMg4.5Mn(G35) with a yield strength of 350 MPa at room temperature. The maximum stress in the collar is 150 MPa. In a test setup with a mechanical load applied to a short stack of collar laminations the measured and computed deflections agreed to within 5%.

increase (decrease) in horizontal radius is coupled with a decrease (increase) in key angle, and these two effects cancel each other almost perfectly. For fields above 6 T the deformed collar touches the very stiff iron yoke and no further deflection occurs.

5.5   Pre-compression of coil and measurement of internal forces

We have seen that the field quality requires a high positional accuracy of the windings of about 20 μm while the quench safety calls for a large internal pre-stress of about 50 MPa. To some approximation, the coil package can be considered as a compressed spring but it is far from being an ideal spring: one observes friction, plastic yield and hysteresis. To ensure that the collars provide the correct geometry and pre-stress after assembly, the coils have to be manufactured with a well-controlled oversize which can be determined only experimentally. A further complication arises from the differential shrinkage of the various materials during cooldown. Between room and liquid helium temperature, the relative shrinking is

coil package $\cong 3.3 - 3.9 \cdot 10^{-3}$,

stainless steel $\cong 3.0 \cdot 10^{-3}$,

aluminium $\cong 4 \cdot 10^{-3}$,

soft iron $\cong 2 \cdot 10^{-3}$.

If the coil is clamped with aluminium the pre-stress should therefore increase slightly upon cool-down whereas with stainless steel collars it should decrease. Soft iron does not appear very adequate as a collaring material since an enormous room temperature pre-stress would be necessary. This may be dangerous for the Kapton insulation which starts to yield at about 70 MPa at room temperature.

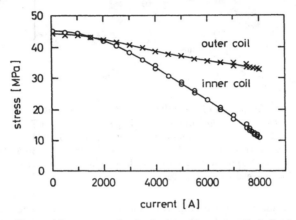

Figure 5.11: Measured pre-stress in the inner and outer coil shell of an SSC model dipole (courtesy C. Goodzeit).

The magnet group at Brookhaven has developed a strain gauge system to measure forces inside the coil during the assembly in the collaring press but also in cold magnets (Goodzeit 1989). As an example we show in Fig. 5.11 the stress in a 1.8-m-long SSC model dipole, measured as a function of current. Starting from 45 MPa at $I = 0$ the pre-stress in the inner coil shell drops more rapidly than in the outer shell due to the larger magnetic forces. But even at 8000 A, far beyond the nominal current of 6600 A, the inner coil shell keeps a positive pre-stress[3].

5.6   Forces between coil and yoke

The coil has to be well centred in the yoke to avoid not only field distortions, but also asymmetry forces between coil and yoke. The right half of the dipole coil is attracted

---

[3]For a few SSC prototype dipoles a complete loss of pre-stress at high field was recorded without premature quenches. While this indicates that some coil motion might be allowed if friction is negligible, a safe design should be based on the requirement that sufficient pre-stress is left over at maximum field.

by the image currents on the right, the left half is pulled to the left. The two forces balance each other in the case of symmetry but the equilibrium is unstable. If the coil is shifted to one side the force in this direction increases whereas the force in the opposite direction decreases. For the Tevatron warm-iron dipole the net force between coil and yoke amounts to 2500 N/m at 4.5 T for a displacement of 0.5 mm. Many supports are needed between coil and yoke to provide a good centring in spite of the asymmetry forces. These supports lead to a rather high heat flux from the warm yoke to the liquid helium vessel. The cold-iron magnet has a clear advantage in this respect: the coil is rigidly centred in the yoke and both are at the same temperature.

5.7 Longitudinal forces

In the coil heads the Lorentz forces act in the longitudinal direction and tend to lengthen the coil. For the HERA dipole the forces add up to about 15 tons at 5 T. The coil itself can bear these forces; it would elongate elastically by about 3 mm. Such an elongation of the coil is undesirable because it changes the field integral and, more dangerously, leads to frictional heating or slip-stick motion between coil and collars or collars and yoke which might trigger quenches. In some of the early 17-m-long SSC prototype magnets premature quenches were observed which could be traced back to a motion induced by the longitudinal forces (Peoples 1989). The best solution is to confine the coil heads by stainless steel end plates which are welded to a longitudinal support structure like the stainless steel tube serving as liquid helium container.

References

H. Brück, R. Meinke and P. Schmüser, *Methods for magnetic measurements of the super-conducting HERA magnets*, Kerntechnik **56** (1991) 248

J.K. Cobb and J.J. Muray, *Magnetic center location in multipole fields*, Nucl. Instr. Meth. **46** (1967) 99

T. Ogitsu et al., *Mechanical performance of 5-cm-aperture, 15-m-long SSC dipole magnet prototypes*, IEEE Trans. **ASC-3** (1993) 686

J. Peoples, *Status of the SSC superconducting magnet program*, IEEE Trans. **MAG-25** (1989) 1444

H. Preissner et al., *A new device for production measurements of field integral and field direction of superconducting dipole magnets*, DESY report HERA 90-07 (1990)

A.V. Tollestrup, *The amateur magnet builder's handbook*, Fermilab report UPC-86, 1979

D. Trbojevic et al., *Alignment and survey of the elements in RHIC*, Proc. Part. Accel. Conf. Dallas 1995

Further reading
J. Buckley et al., *Mechanical behaviour during excitation of the first CERN 10 T dipole model magnet for LHC*, Europ. Acc. Conf. London 1994, World Scientific 1994,

  p. 2286

R. Hanft et al., *Magnetic field properties of Fermilab Energy Saver dipoles*, IEEE Trans.
   Nucl. Sci. **NS-30** (1983) 3381

E.E. Schmidt et al., *Magnetic field data on Fermilab Energy Saver quadrupoles*, IEEE
   Trans. **NS-30** (1983) 3384

P. Schmüser, *Field quality issues in superconducting magnets*, Proc. Part. Accel. Conf.
   San Francisco 1991, p. 37

P. Schmüser, *Properties and practical performance of sc magnets in accelerators*, Proc.
   Third Eur. Part. Acc. Conf. Berlin 1992, Edition Frontières (1992) 284

N. Siegel et al., *Mechanical tests on the prototype LHC lattice quadrupoles*, IEEE Trans.
   MAG-30 (1994) 2475

F. Turkot et al., *Maximum field capability of Energy Saver superconducting magnets*, IEEE
   Trans. **NS-30** (1983) 3387

Chapter 6

Persistent Currents

6.1 Theoretical model of persistent magnetization currents

6.1.1 Superconductor magnetization

Persistent magnetization currents in the superconductor are the source of severe field distortions at low excitation of a superconducting accelerator magnet. These bipolar currents generate all multipoles which are allowed by coil symmetry: b_1, b_3, b_5, b_7, ... in a dipole, b_2, b_6, b_{10}, b_{14}, ... in a quadrupole. A distinct hysteresis behaviour is observed: the multipole fields have opposite signs for increasing and decreasing main field, respectively.

We now proceed to construct a theoretical model. When the field in a magnet is changed, currents are induced in all conducting materials exposed to the varying magnetic flux, in particular of course in the superconducting cable. Most superconducting accelerator magnets are made from the Rutherford type cable shown in Fig. 3.9. The HERA dipole cable, for instance, consists of 24 wires ('strands') with a diameter of 0.84 mm. Each strand contains 1230 NbTi filaments of 14 μm diameter, embedded in a copper matrix. To reduce induction effects the strands in the cable are transposed with a pitch length of 95 mm and the filaments in the strands are twisted with a twist length of 25 mm.

We can distinguish three types of induced currents in the coil: eddy currents between different strands in the cable, coupling currents between different filaments inside a strand, and finally the so-called *magnetization currents* inside individual filaments. The cable and strand eddy currents will be treated in Chap. 7. Both are found to decay exponentially with time constants below a second although there are certain types of long-lived eddy currents too (see Sect. 7.3).

Truly persistent currents exist only within individual filaments. Their computation is based on the experimentally verified critical state model (see Chap. 2.4.5). According to this model a hard superconductor tries to expel any external field change by generating a bipolar current distribution with the highest possible density, namely the critical current density $J_c(B, T)$ at the given local field and temperature. Let us study the response of a superconductor filament to a homogeneous external field B_e which is first raised from zero and then lowered again. With increasing B_e, a $\cos \phi$-like current distribution is induced (Fig. 6.1a) producing a homogeneous shielding field B_s which just cancels B_e in the current-free region of the filament. Following Wilson (1983) we approximate the boundary of the current-free region by an ellipse with large half axis $a = r_f$ (filament radius), small half axis b and eccentricity $\varepsilon = \sqrt{1 - (b/a)^2}$. The contribution of the two area elements $dxdy$ at the locations $(+x, y)$ and $(-x, y)$ to the shielding field is

$$dB_s = -2 \frac{\mu_0 J_c dxdy}{2\pi r} \cos \phi$$

with $r = \sqrt{x^2 + y^2}$ and $\cos \phi = x/r$. The shielding field at $r = 0$ is found by integration

$$B_s = -\frac{\mu_0 J_c}{\pi} \int_{-a}^{a} \left[\int_{u(y)}^{v(y)} \frac{x}{x^2 + y^2} dx \right] dy .$$

The lower limit $u(y) = b\sqrt{1 - y^2/a^2}$ of the x integral derives from the equation of the ellipse $x^2/b^2 + y^2/a^2 = 1$, the upper limit $v(y) = \sqrt{a^2 - y^2}$ from the circle $x^2/a^2 + y^2/a^2 = 1$. The integrations can be performed analytically.

$$B_s = -\frac{2\mu_0 J_c r_f}{\pi} \left(1 - \sqrt{1 - \varepsilon^2} \cdot \frac{\arcsin \varepsilon}{\varepsilon} \right) . \tag{6.1}$$

The highest field which can be shielded from the interior of the filament is called the 'penetrating' field B_p and is obtained for an ellipse shrunk to a line, i.e. $\varepsilon = 1$.

$$B_p = \frac{2\mu_0 J_c r_f}{\pi} . \tag{6.2}$$

Figure 6.1b shows the currents in the 'fully penetrated' filament. The applied field may be raised to much larger values than B_p which is only about 0.13 T for the HERA conductor. In that case the same current pattern is obtained as in Fig. 6.1b but with a non-vanishing field throughout the filament. If now the field is decreased again, persistent currents with opposite polarity are superimposed because the superconductor tries to avoid a change of the inner field. A more complicated current pattern arises as indicated in Fig. 6.1c. The current loops are assumed to be closed at the coil ends. The effect of the short coil ends on the integrated multipole fields can be neglected.

The magnetic moment m_f of the filament can also be explicitly calculated. Consider again the two area elements $dxdy$ at $(+x, y)$ and $(-x, y)$. They constitute a current loop whose magnetic moment is

$$dm_f = -J_c \, dxdy \cdot 2x \cdot l_f$$

where l_f is the length of the filament. This expression has to be integrated over the area occupied by the positive current.

$$m_f = -2J_c l_f \int_{-a}^{a} \left[\int_{u(y)}^{v(y)} x\,dx \right] dy .$$

The result is (using $a = r_f$ and $\varepsilon^2 = 1 - b^2/a^2$)

$$m_f = -\frac{4}{3} J_c \varepsilon^2 r_f^{\,3} l_f .$$

The magnetization, defined as the magnetic moment per unit volume, is obtained by dividing m_f by the volume $\pi r_f^2 l_f$ of the filament

$$M = -\frac{4}{3\pi} J_c r_f \, \varepsilon^2 \quad . \tag{6.3}$$

To compute M as a function of the external field B_e we first express the eccentricity ε in terms of B_e by using the condition $B_s = -B_e$ and Eq. (6.1). The magnetization assumes its peak value for the fully penetrated filament (Fig. 6.1b):

$$M_p = |M|_{max} = \frac{4}{3\pi} J_c r_f \quad . \tag{6.4}$$

Note that the quantity M_p is not a constant but decreases proportional to the critical current density $J_c(B_e, T)$ when the external field is raised beyond the penetrating field.

Up to now we have disregarded the field-generating *transport current* I_t. Its influence shall be studied for the case of a saturated filament, i.e. $B > B_p$. According to the critical-state model, the transport current must also flow with the critical density J_c but is confined to a small elliptical region at the centre of the filament, see Fig. 6.1d. If we define the average transport current density J_t as transport current per filament I_t, divided by filament area $\pi a^2 = \pi r_f^2$, the small half axis b of the transport-current ellipse is given by the condition

$$I_t = J_t \pi a^2 = J_c \pi ab \quad \Rightarrow \quad b/a = J_t/J_c .$$

This implies that for computing the magnetization Eq. (6.3) must be used with $\varepsilon^2 = 1 - (b/a)^2 = 1 - (J_t/J_c)^2$, so effectively the transport current reduces the magnetization (6.4) of an otherwise saturated filament by the factor $(1 - (J_t/J_c)^2)$. The correction is negligible near the injection field where $J_t \ll J_c$ but leads to a significant reduction of filament magnetization at high excitation of the magnet.

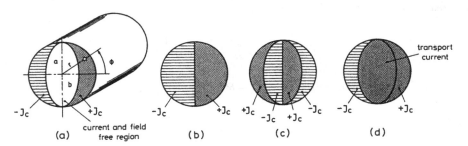

Figure 6.1: Schematic view of the persistent currents which are induced in a superconducting filament by a varying external field. (a) The external field is raised from zero to a value B_e less than the penetrating field B_p. (b) A 'fully-penetrated' filament, i.e. $B_e \geq B_p$. (c) Current distribution which results when the external field is first increased from zero to a value above B_p and then decreased again. (d) Same as (b) but with a large transport current.

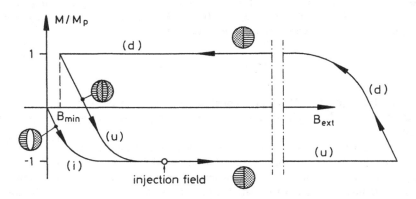

Figure 6.2: The normalized magnetization M/M_p of a NbTi filament as a function of the external field. (i): initial curve, (u): up-ramp branch, (d): down-ramp branch. Also shown are the current distributions in the filament. The field dependence of J_c has been neglected.

From the equations (6.1) and (6.3) one can compute the magnetization as a function of the external field. The result is plotted schematically in Fig. 6.2. We observe a hysteresis behaviour with three different states: Starting at the virgin state the magnetization follows an initial curve (i) and reaches its peak value at $B_e = B_p$. After going up to high field the ramp direction is reversed and M follows the 'down-ramp' branch (d). At a certain minimum current the field is increased again and the magnetization follows the 'up-ramp' branch (u) which has the remarkable feature that M changes its sign from positive to negative values. This is exactly what is observed in the 6-pole and 12-pole coefficients (see Fig. 6.5 below). In Fig. 6.2 also the current pattern in the filament is indicated at different positions of the hysteresis loop. An experimental magnetization curve is presented in Fig. 6.3.

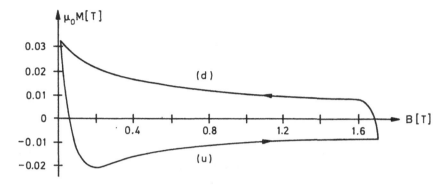

Figure 6.3: The measured magnetization of the HERA dipole conductor as a function of the external field (Ghosh et al. 1985). Note that here the magnetization has not been normalized to its peak value. The width of the hysteresis curve decreases towards higher fields owing to the field dependence of the critical current density.

6.1.2   Calculation of persistent-current multipole fields and comparison with data

The field distortions from persistent currents have been computed by various authors (Green 1971, Duchateau 1972). In the following we describe a program (Brück, Meinke, Müller, Schmüser 1989) developed at DESY to model the persistent-current multipoles of the HERA magnets. In the first step the local field is calculated at any current conductor inside the coil of the magnet. Using the formulae in Sect. 6.1.1 the magnetization currents are then computed from the time variation of the local field following the complete history. In the third step the fields generated by the bipolar or even more complicated current patterns are determined. The field calculation will be described for a dipole coil. We start with four symmetrically arranged filaments in the dipole coil in which bipolar currents have been induced by the increasing main field (see Fig. 6.4a). The current distribution in each filament can be replaced by a pair of line currents $+I$ and $-I$ whose strength equals the integrated current density and whose separation d is chosen such that the computed filament magnetization m_f is obtained (for the fully penetrated filament, $I = J_c \pi r_f^2 / 2$ and $d = 8 r_f / (3\pi)$). Since $d \ll R$, the vector potential of the four current pairs in Fig. 6.4b can be derived by first-order Taylor expansion from the vector potential A of four single currents (see Eq. (4.15)):

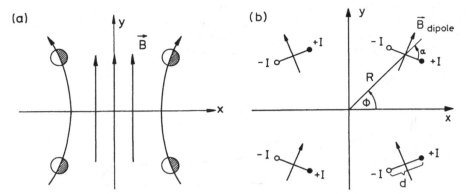

Figure 6.4: (a) Magnetization currents induced by the time-varying main field in four symmetrically arranged filaments inside the dipole coil. (b) Equivalent pairs of line currents. The separation d between the positive and negative currents is grossly exaggerated.

$$A^{pair} = \frac{\partial A}{\partial R}\Delta R + \frac{\partial A}{\partial \phi}\Delta \phi \quad . \tag{6.5}$$

With the relations $\Delta R = d\cos\alpha$, $\Delta\phi = -d\sin\alpha/R$ we obtain

$$A^{pair}(r,\theta) = -\frac{2\mu_0 I d}{\pi R}\sum_{n=1,3,\ldots}\left(\frac{r}{R}\right)^n\cos(n\theta)\cos(n\phi+\alpha) \quad . \tag{6.6}$$

The influence of the iron yoke with an inner bore radius R_y is taken into account by the image current method. The image of a current pair at a radius R and angle ϕ appears at $R' = R_y^2/R$ and $\phi' = \phi$. The separation of the image currents is $d' = d \cdot R'/R$ and the angle with respect to the position vector $\vec{R'}$ is $\alpha' = \pi - \alpha$. Replacing the quantities R, d, α in Eq. (6.6) by R', d' and α' one gets the iron contribution A'^{pair} to the vector potential. The resulting multipole expansion of the azimuthal field component is then given by

$$B_\theta(r,\theta) = -\frac{\partial}{\partial r}(A^{pair} + A'^{pair})$$

$$B_\theta(r,\theta) = \frac{2\mu_0(I\cdot d)}{\pi R^2}\sum_{n=1,3,\ldots} n\cos(n\theta)$$

$$\cdot\left[\left(\frac{r}{R}\right)^{n-1}\cos(n\phi+\alpha) - \frac{R}{R'}\left(\frac{r}{R'}\right)^{n-1}\cos(n\phi-\alpha)\right] \quad . \tag{6.7}$$

For the product $(I\cdot d)$ we insert the magnetic moment per unit length derived from Eq. (6.3). Expression (6.7) has to be summed over all NbTi filaments in one quarter of the dipole coil and divided by the main dipole field to obtain the multipole coefficients.

With Eq. (6.7) we have derived a remarkable result: the persistent currents generate the same 'allowed' multipoles as the transport current, namely the normal multipoles b_n of the orders $n = 1, 3, 5, \ldots$. The even orders $n = 2, 4, \ldots$ are absent and no skew multipoles a_n appear. In a quadrupole coil, correspondingly, the persistent currents obey again the same symmetries as the transport current and generate only the allowed normal multipoles b_2, b_6, b_{10}, b_{14}, $\ldots$. 'Unallowed' poles can only be present when the superconductor properties, for instance the critical current density, are not uniform over the coil cross section.

An important ingredient to the model is the critical current density $J_c(B,T)$ at low fields which is unfortunately not easily accessible. The manufacturers of superconducting cables measure critical currents usually at fields of 5 – 6 T. From magnetization measurements at low fields (Fig. 6.3) one deduces J_c by making use of expression (6.4) and correcting for the volume fraction of superconductor in the cable. The uncertainty in the critical current at low fields is about 10% and this is the dominant uncertainty in the calculation of persistent current multipoles. The filament diameter is known to about 5%. A useful parametrization of $J_c(B,T)$ can be found in (Green 1989).

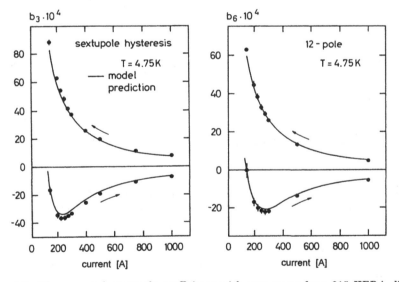

Figure 6.5: The averaged sextupole coefficients with rms errors from 315 HERA dipoles (12-pole coefficients from 236 quadrupoles) as a function of coil current. The multipole coefficients b_n (a_n) are defined as the ratio of the corresponding multipole fields B_n (A_n) to the main field (dipole B_1, quadrupole B_2) at the reference radius $r_0 = 25$ mm. The curves represent the model predictions. The ramp direction of the current is indicated by arrows. Before starting the measurements, a current cycle 50 A $\rightarrow$ 6000 A $\rightarrow$ 50 A was performed to establish a well-defined initial condition for the superconductor magnetization.

The averaged sextupole data of 315 HERA dipoles and the 12-pole data of 236 quadrupoles are shown in Fig. 6.5 for increasing and decreasing main field[1]. The predictions of the model, shown as continuous curves, are in excellent agreement with the measurements.

The persistent currents have also a significant influence on the main dipole field and quadrupole gradient. Their contribution is denoted by $\tilde{B}_1$ resp. $\tilde{g}$ and is plotted in Fig. 6.6. Again a hysteresis curve is observed and the data are in fairly good agreement with the model prediction[2].

At the HERA injection energy the main dipole field (quadrupole gradient) is 0.5% (0.2%) lower than the value computed from the coil current. Of course a correction is needed to match HERA to the energy of the pre-accelerator.

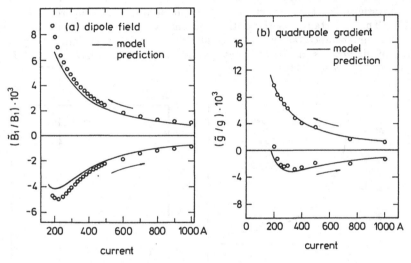

Figure 6.6: (a) The relative contribution $\tilde{B}_1/B_1$ of the persistent currents to the main dipole field. (b) Relative contribution $\tilde{g}/g$ to the main quadrupole gradient. Solid curves: model prediction.

The superconductor magnetization and the resulting multipoles are proportional to the filament diameter and the critical current density. At low field J_c is almost an

---

[1]In Fig. 6.5 only the contribution of the persistent currents to the multipoles is plotted. The fairly large geometric sextupoles in the dipoles were determined at 3000 A and subtracted on a magnet-by-magnet basis from the 250 A data. A similar procedure was applied in the quadrupoles.

[2]It should be mentioned that the dipole field was measured with an NMR probe of a few mm size. The $\tilde{B}_1$ data are therefore affected by the longitudinal periodicity discussed in Sect. 6.3 while the sextupole measurements were carried out with 2.40-m-long pick-up coils which average over the oscillation. In addition there is a small contribution to the dipole field from the remanence of the iron yoke.

order of magnitude larger than at 5 Tesla so the multipoles are particularly worrysome at the injection energy of the accelerator. Of course, nobody wants to sacrifice a high J_c just to reduce the undesirable persistent-current effects but a reduction in filament size is certainly advisable. The cable in the HERA dipoles has fairly thick filaments ($14-16$ μm). When the HERA superconductor was specified in 1984, an optimization of costs and critical current resulted in a filament diameter above 10 μm. In the past years great progress has been made towards finer filaments. For the SSC magnets, a diameter of 6 μm had been foreseen, a similar number is envisaged for LHC.

6.2 Time dependence of persistent-current effects

A time dependence of the persistent-current effects was first observed at the Tevatron (Finley et al. 1987). When the accelerator was converted into a proton-antiproton collider chromaticity shifts were observed during the one-hour injection time of antiprotons at 150 GeV which caused a certain beam loss[3]. Measurements on spare dipoles revealed that the reason was a time-variation of the persistent-current sextupole. The data, plotted in Fig. 6.7, could not be described in terms of a single exponential but required two or more time constants.

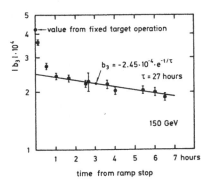

Figure 6.7: Observed sextupole drift in a Tevatron dipole with exponential fit for times above 1 hour. (© 1987 IEEE)

Similar observations were made on pre-series HERA magnets and it was found out (Mess, Schmüser 1988) that the multipole fields are much better described by a logarithmic instead of an exponential time dependence. As an example we show in Fig. 6.8 the time dependence of the dipole and sextupole components in a dipole magnet (a similar behaviour is found for the 12-pole in the quadrupoles). A sextupole

---

[3]Chromaticity is the dependence of betatron oscillation frequency on particle momentum, see Chap. 9.1.

drift with $\log t$ was also seen in Tevatron magnets (Herrup et al. 1989, Hanft et al. 1989).

The injection of 40 GeV protons into HERA starts about 10 minutes after the injection field of 0.227 T (corresponding to a magnet current of 245 A) has been established and may take some 20 minutes. During this time interval the drift is well represented by the form $A - R \log t$. The slope R will be called the logarithmic decay rate (compare also Eq. (2.16)) and is identical with the change of the multipole fields per decade of time, for instance between $t = 200$ s and $t = 2000$ s. The logarithmic decay rates R_1 of the dipole field and R_3 of the sextupole components are plotted against each other in Fig. 6.9 for more than 200 magnets. The time measurements were performed after an initial current cycle

$$0 \text{ A} \rightarrow 6000 \text{ A} \rightarrow 50 \text{ A} \rightarrow 250 \text{ A}$$

in which the maximum current I_{max} was chosen 1000 A above the nominal operating current in HERA. The decay rates of the sextupole and dipole components are strongly correlated indicating a common origin for both decays. The data are compatible with the assumption of a logarithmically decreasing critical current density, suggesting thermally-activated flux creep as the underlying physical mechanism.

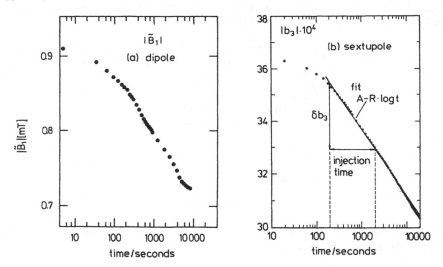

Figure 6.8: Time dependence of the persistent-current field distortions in a HERA dipole at a field of 0.227 T: (a) absolute value of the contribution $\tilde{B}_1$ to the dipole field, (b) sextupole component.

There are, however, several puzzling observations which do not fit into this picture:

- The decay rates measured in magnets are usually much larger than those in cable samples (Ghosh 1990, Collings 1990).

- There is a considerable variation from magnet to magnet and in particular between magnets made from superconducting cables of different origin.

- The data plotted in Fig. 6.10a demonstrate that the persistent-current decay rate at low field is strongly influenced by the maximum excitation B_{max} of the magnet in the preceding field cycle. In contrast to this, B_{max} has no measurable effect on the shape and width of the sextupole hysteresis curve at low fields. Also in the model discussed in Sect. 6.1 the superconductor magnetization at low field, say 0.23 T, is the same for $B_{max}=0.7$ T or $B_{max}=4.7$ T.

In order to clarify the role of flux creep in accelerator magnets, an experimental setup was devised at DESY permitting measurement of the average magnetization of a 5-m-long sample of the insulated HERA dipole cable as well as its time dependence. (Usually magnetization measurements are done on samples of a few mm length so possible effects of the cable pitch cannot be observed.) The results are plotted in Fig. 6.10b. The magnetization decay at low field (actually $B = 0$ in this case) is less than 1% per decade of time but is totally independent of the maximum field B_{max} in the preceding cycle. The observed decay rate agrees well with other data on flux creep in NbTi.

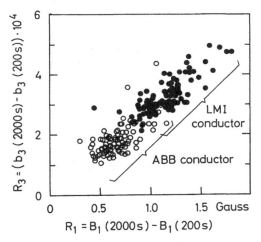

Figure 6.9: Correlation between the logarithmic decay rates of the dipole and sextupole components (Brück, Jiao et al. 1989).

Figure 6.11a, taken from Devred, DiMarco et al. (1991), indicates that the sextupole time decay exhibits two phases: (1) in the first 300 seconds the decay rate is small and independent of the maximum current I_{max} in the pre-cycle; (2) at larger times the decay rate increases almost linearly with I_{max}. For $I_{max} < 2000$ A, or when the pre-cycle is omitted altogether, the observed sextupole decay is compatible with flux creep. At larger I_{max} obviously a new mechanism comes in. The two phases are also visible in the HERA data of Fig. 6.8.

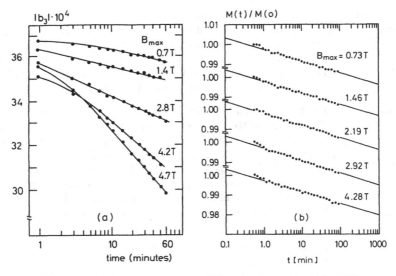

Figure 6.10: (a) Decay of the sextupole in a HERA dipole at 0.23 T for different values of the maximum field in the initializing cycle $0 \to B_{max} \to 0.04$ T$\to 0.23$ T (Brück, Jiao et al. 1989). (b) Magnetization decay at zero field in a long sample of HERA cable for different maximum fields in the initial field cycle $0 \to B_{max} \to 0$ (Halemeyer et al. 1993).

Interesting additional information comes from studies of the temperature dependence. Cross and Goldfarb (1991) have measured the magnetization decay in a NbTi sample at 3.5 and 4.0 K. The ratio $M(t)/M(0)$ is described by the same function of time for both temperatures (Fig. 6.11b). In the model of thermally activated flux creep one expects a linear rise of the creep rate $R_0 = dM/d\ln t$ with temperature T, see Eq. (2.16), but the initial magnetization $M(0)$ is roughly proportional to $1/T$, hence the T dependence should drop out in the ratio $M(t)/M(0)$. On the other hand, if the temperature is lowered from 4.0 to 3.5 K during the experiment, a reduced creep rate is expected because the initial magnetization corresponds to the critical current density at 4.0 K and is lower than the critical magnetization at 3.5 K. The exponential dependence in Eq. (2.14) leads therefore to a markedly reduced creep rate which is indeed observed. The opposite effect happens when the field sweep is performed at 3.5 K and then the temperature in raised to 4.0 K: the decay rate increases by more than a factor of two. A similar experiment was performed with a 17-m-long SSC magnet at 3.8 K and 4.35 K. The data without pre-cycle are shown in Fig. 6.12a. The 3.8 K curve is simply parallel-shifted with respect to the 4.35 K curve because of the higher critical current density at 3.8 K. The decay rates are nearly the same.

With an initial field sweep at 4.35 K and a temperature reduction to 3.8 K after 600 seconds the sextupole decay completely stops.

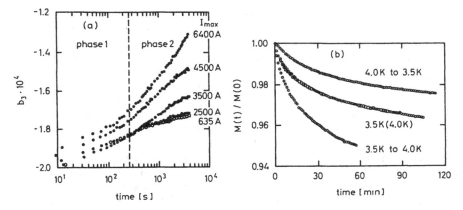

Figure 6.11: (a) Influence of the maximum current in the pre-cycle on the time decay of the sextupole in a 4-cm-aperture 17-m-long SSC dipole prototype. Sequence: cleansing quench, pre-cycle $0 \rightarrow I_{max} \rightarrow 120$ A $\rightarrow 635$ A. The maximum current was kept for 1 hour. The decay was measured at 635 A, corresponding to the SSC injection energy (Devred, DiMarco et al. 1991). (b) Magnetization decay studies of a NbTi sample: decay at constant temperature of either 3.5 or 4.0 K (middle curve), with a temperature decrease during the measurement from 4.0 to 3.5 K (top curve) and, finally, with a temperature increase from 3.5 to 4.0 K (bottom curve). (Cross, Goldfarb 1991). (a: © 1991 IEEE)

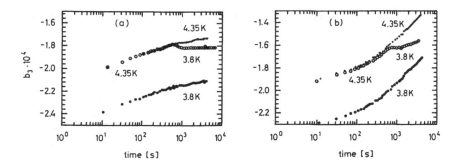

Figure 6.12: Sextupole decay in an SSC dipole at constant temperature ($T = 4.35$ K or $T = 3.8$ K). A third measurement was started at 4.35 K and after 600 seconds the temperature was lowered by 0.5 K. (a) Measurement without pre-cycle: quench, single ramp $0 \rightarrow 635$ A. (b) Measurement with large pre-cycle: quench, $0 \rightarrow 6400$ A $\rightarrow 120$ A $\rightarrow 635$ A (Devred, DiMarco et al 1991). (© 1991 IEEE)

This is in qualitative agreement with the flux creep data of Cross and Goldfarb. Quite a different behaviour is observed when the pre-cycle is extended to 6400 A (Fig. 6.12b). The curves taken at constant temperatures of 4.35 and 3.8 K are again almost parallel; however when the temperature is lowered from 4.35 to 3.8 K during the measurement the sextupole decay is not stopped but continues and gradually approaches its original rate.

All these observations demonstrate that flux creep alone cannot account for the strong time dependence of persistent-current fields in magnets. Now there is an essential difference between the superconductor in a test sample and that in the coil, namely the fact that the time derivative $\dot{B}$ of the field penetrating the cable is not constant in a magnet coil but varies along the length of the conductor. In the coil heads, for instance, it may be considerably smaller than in the straight section. We will show in Chap. 7 that a longitudinal variation of $\dot{B}$ induces eddy currents in the strands which decay with long time constants. Effectively this is a redistribution of current among the strands in the cable. It has been conjectured[4] that a current redistribution may affect the superconductor magnetization due to the associated changes in the local magnetic field. We will come back to this question in Sect. 7.3.

Another difference between the superconductor sample and the cable in the coil is the transport current that generates the field. Cross and Goldfarb (1991) have studied the influence of a transport current in an SSC cable sample on the superconductor magnetization and its decay. For transport currents below 20% of the critical current no change in decay rate was observed. This implies that the relatively small transport current at the injection field of accelerator magnets cannot be responsible for the large time dependencies.

6.3   Longitudinal periodicity of persistent-current fields

In an attempt to measure the sextupole decay at various positions in a HERA dipole, using a three-Hall-probe detector with good spatial resolution, the surprising discovery was made (Brück, Gall et al. 1991) that the sextupole field exhibits a pronounced periodic pattern along the axis of the magnet. The first evidence for this unexpected effect is shown in Fig. 6.13, curve (a). The magnet was subjected to a current cycle typical for accelerator operation ($0 \rightarrow 5500$ A $\rightarrow 50$ A $\rightarrow 250$ A) and then kept at 250 A for one hour to eliminate most of the logarithmic time dependence. An almost sinusoidal structure was observed with a remarkably large amplitude. After going through another cycle 250 A $\rightarrow 2000$ A $\rightarrow 250$ A curve (b) was obtained and lowering the current to 100 A finally curve (c). The oscillatory pattern is practically the same in all three cases, concerning amplitude, wavelength and phase, whereas

---

[4]R. Stiening, private communication.

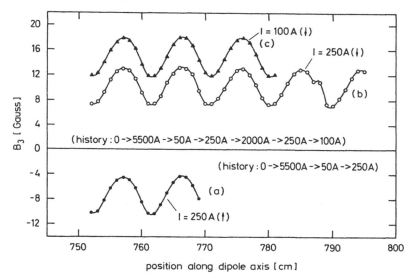

Figure 6.13: Sextupole field B_3 at $r_0 = 25$ mm as a function of the coordinate z along the axis of a HERA dipole. Curve (a): measurement at 250 A on the 'up-ramp' branch of the hysteresis curve, following an initial current cycle $0 \to 5500$ A $\to 50$ A $\to 250$ A. Curves (b), (c): at 250 A resp. 100 A on the 'down-ramp' branch (Brück, Gall et al. 1991).

the average values of the sextupole lie on the standard hysteresis curve[5]. After the power supply was switched off the oscillatory pattern persisted without attenuation for more than 12 hours but vanished immediately when the magnet was warmed up to 20 K. This proves that the effect is associated with the superconducting state. Another interesting result is plotted in Fig. 6.14: in a 'virgin' magnet, immediately after cooldown, the oscillation is unvisible at very low excitation but appears only gradually with increasing coil current. The oscillation is particularly pronounced at high fields, as can be seen from Fig. 6.15 where the sextupole in an SSC dipole (serial number DCA 213) is plotted for magnet currents of 650 and 7000 A (Ghosh, Robins, Sampson 1993). At high field the oscillation amplitude is much larger than the average value of the sextupole field. It should be noted that the detailed behaviour is not representative for other magnets except for the general trend that the oscillation becomes more pronounced at higher excitation of the magnets.

A clue to an understanding of the longitudinal periodicity is the observation that the wavelength of 94 mm in the HERA magnets is in close agreement with the transposition pitch of 95 ± 2 mm in the Rutherford cable. These findings were confirmed at BNL, SSCL and CERN, and also in those magnets the period agreed with the

---

[5]The standard multipole measurements were made with rotating pick-up coils of 1.10 to 2.40 m length which average over the oscillation.

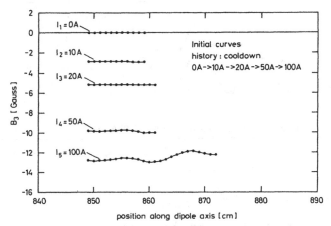

Figure 6.14: Sextupole field at $r_0 = 25$ mm as a function of z for the initial excitation of a previously quenched magnet. (Brück, Gall et al. 1991).

cable pitch. The periodic pattern can be qualitatively explained by assuming a current imbalance between the strands of the cable. The simplest model is that of a two-strand cable in which one wire carries a higher current than the other. This is sketched schematically in Fig. 6.16. The resulting zig-zag pattern of currents leads to a sequence of alternating magnetic moments and to an almost sinusoidal field perturbation along the axis. The origin of the current imbalance will be analyzed in Sect. 7.3.

In Fig. 6.17 we show data from a HERA quadrupole taken with a 20-mm-long rotating pick-up coil. Not only the allowed multipole fields B_2, B_6, ... are found to be modulated but all A_n and B_n exhibit the periodic pattern. This proves that the currents responsible for this pattern do not obey any coil symmetries. Note that the unallowed multipoles average to zero.

It is worth mentioning that the oscillating multipole fields should have a negligible effect on the proton beam emittance[6] since their wavelength is orders of magnitude smaller that the betatron wavelength.

---

[6]The emittance is the area in phase space occupied by a particle beam, see Chap. 9.1.

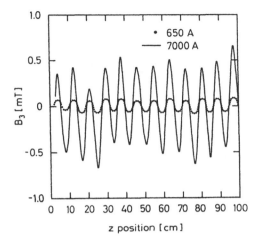

Figure 6.15: Longitudinal periodicity of sextupole in an SSC dipole at 650 A and 7000 A (Ghosh, Robins, Sampson 1993). (© 1993 IEEE)

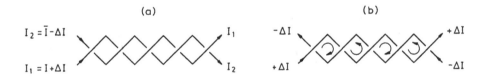

Figure 6.16: (a) Simplest model for the generation of longitudinally periodic field perturbations. A two-strand Rutherford cable is considered in which one stand carries more current than the other. (b) Subtracting the average transport current, a bipolar current pattern is obtained which is equivalent to a periodic sequence of alternating magnetic moments.

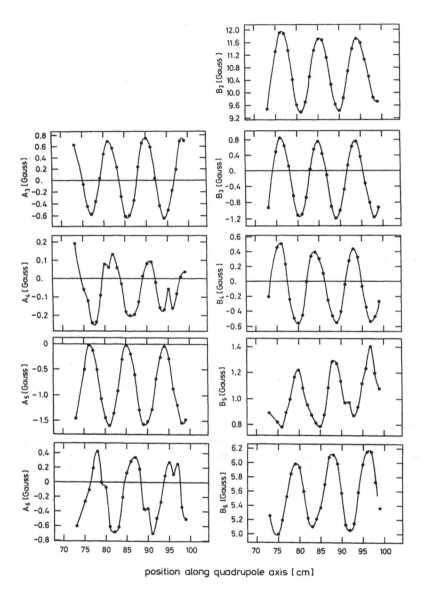

Figure 6.17: Periodicity of the skew (A_n) and normal (B_n) multipole fields at $r_0 = 25$ mm in a HERA quadrupole. The data were taken for coil current $I = 0$, following a current cycle $0 \rightarrow 6000$ A $\rightarrow 0$. Note the large average values of the allowed multipole fields B_2 and B_6.

References

H. Brück, R. Meinke, F. Müller, P. Schmüser, *Field distortions from persistent currents in the superconducting HERA magnets*, Z. Physik **C44** (1989) 385

H. Brück , Z. Jiao et al., *Time dependence of persistent current effects in the superconducting HERA magnets*, 11th Int. Conf. Magn. Techn. MT-11, Tsukuba, Japan 1989, p. 141

H. Brück, D. Gall et al., *Observation of a periodic pattern in the persistent-current fields of the superconducting HERA magnets*, Proc. IEEE Particle Accelerator Conference, San Francisco 1991, p. 2149

R.W. Cross, R.B. Goldfarb, *Enhanced flux creep in Nb-Ti superconductors after an increase in temperature*, Appl. Phys. Lett. **58** (1991) 415

R.W. Cross, R.B. Goldfarb, *Hall probe magnetometer for SSC magnet cables: effect of transport current on magnetization and flux creep*, IEEE Trans. **MAG-27** (1991) 1796

A. Devred, J. DiMarco et al., *Time decay measurements of the sextupole component of the magnetic field in a 4-cm-aperture, 17-m-long SSC dipole magnet prototype*, IEEE Part. Acc. Conf., San Francisco 1991, p. 2480

J.-L. Duchateau, *Etude du champ remanent dans les aimants multipolaires supraconducteurs*, Department Saturne internal report SEDAP/72-109 (1972)

D.A. Finley et al., *Time dependent chromaticity changes in the Tevatron*, Proc. IEEE Part. Acc. Conf., Washington D.C. 1987, p. 151

A.K. Ghosh et al., IEEE Trans. **MAG-21** (1985) 328 and A.K. Ghosh, private communication

A.K. Ghosh, K.E. Robins, W.B. Sampson, *Axial variations in the magnetic field of superconducting dipoles and quadrupoles*, IEEE Part. Acc. Conf., Washington D.C. 1993

M.A. Green, *Residual fields in superconducting dipole and quadrupole magnets*, IEEE Trans. **NS-18** (1971) 664

M.A. Green, *Calculating the J_c, B, T surface for niobium-titanium using a reduced-state model*, IEEE Trans. **MAG-25** (1989) 2119

M. Halemeyer et al., *A new method for determining the magnetization of superconducting cables and its time dependence*, IEEE Trans. **ASC-3** (1993) 168

R.W. Hanft et al., *Studies of time dependence of fields in Tevatron superconducting dipole magnets*, IEEE Trans. **MAG-25** (1989) 1647

D.A. Herrup et al., *Time variations of fields in superconducting magnets and their effects on accelerators*, IEEE Trans. **MAG-25** (1989) 1643

K.-H. Mess, P. Schmüser, *Superconducting accelerator magnets*, Lectures at the CERN-DESY School 'Superconductivity in Particle Accelerators', Hamburg 1988, CERN report 89-04 (1989)

M.N. Wilson, *Superconducting Magnets*, Chap. 8, Clarendon Press, Oxford 1983

Further reading

M. Ashkin, *Flux distribution and hysteresis loss in a round superconducting wire for the complete range of flux penetration*, J. Appl. Phys. **50** (1979) 7060

J. Chaussy et al., *Flux creep phenomenon in multifilamentary superconducting wires*, Phys. Lett. **87A** (1981) 61

K.F. Gertsev et al., *Dynamic effects in the magnetic field of superconducting dipoles for UNK*, IEEE Trans. **MAG-24** (1988) 812

W.S. Gilbert et al., *The effect of flux creep on the magnetization field in the SSC dipole magnets*, Adv. Cryog. Eng. **36A** (1990) 233

M. Kuchnir and A.V. Tollestrup, *Flux creep in a Tevatron cable* (1988)

R. Wolf, *Persistent currents in LHC magnets*, IEEE Trans. **MAG-28** (1992) 374

Chapter 7

Eddy Current Effects in Superconducting Magnets

Most superconducting accelerator magnets are made from the Rutherford-type cable described in Chap. 3. The cable consists of 20 to 40 strands of 0.7 to 1.3 mm diameter which are twisted around each other and shaped into a two-layer flat cable. Usually the cable is compressed to a trapezoidal cross section and its average thickness is less than twice the strand diameter. At their cross-over points, the strands are indented and have a fairly large contact area which, in combination with the high pre-stress in the collared coil, leads to small inter-strand resistances of a few $\mu\Omega$. The two layers of strands thus form an arrangement of loops in which eddy currents of sizeable strength can be induced by a time-varying magnetic field. These eddy currents between different strands will be referred to as *cable eddy currents*. Another type are the *coupling currents* between different filaments inside a strand. Additional eddy currents arise in the copper wedges of the coil, in the collars, the yoke and other conducting materials.

7.1   Coupling currents within a strand

The filaments inside a strand are strongly coupled through the copper matrix. An effective means to reduce the filament coupling is provided by twisting the wire. Thereby the length of the loop that is exposed to the time-varying magnetic field is considerably shortened and the contributions from adjacent loops alternate in sign. In a multifilamentary twisted superconducting wire, the eddy currents run in a zig-zag fashion along the wire. The path is partly superconductive (inside the filaments), partly normal (in the copper matrix). A very clear analysis of the problem can be found in the book by Wilson (1983). The screening currents obey a $\cos\phi$ dependence and thus produce a homogeneous shielding field B_s. The inner field $B_i = B_e - B_s$ is also homogeneous. In case of a constant ramp rate, the strand magnetization resulting from the eddy currents is found to be

$$M_s = 2\dot{B}_i\tau/\mu_0 \,. \tag{7.1}$$

Here, τ is a time constant that is related to the twist pitch length l_{twist} of the strand and the effective transverse resistivity ρ_t of the copper-NbTi composite:

$$\tau = \frac{\mu_0}{2\rho_t} \left(\frac{l_{twist}}{2\pi} \right)^2 . \tag{7.2}$$

A typical value for the twist pitch is 25 mm. The dissipated power P per unit volume is derived from the formula

$$P dt = M_s dB_i = \frac{2\dot{B}_i \tau dB_i}{\mu_0} = \frac{2\dot{B}_i^2 \tau}{\mu_0} dt ,$$

from which follows

$$P = \frac{2\dot{B}_i^2 \tau}{\mu_0} . \tag{7.3}$$

Additional intra-strand eddy currents arise in the central copper core and in the outer copper sheath of the strand. This has been analyzed by Devred and Ogitsu (1994). A general treatment of intra-strand eddy currents is found in (Mulder, Niessen 1993).

The overall time constant τ has been estimated to be in the order of 5 ms for the SSC cable. Measurements of a.c. (alternating current) losses on LHC cables yield time constants around 20 ms (Verweij 1995). So the strand magnetization resulting from the intra-strand eddy currents is quite short-lived and its contribution to the field distortions during the acceleration of the particle beam is found to be negligible in comparison with that of the persistent-currents. The ramp time T_{ramp} from zero to the maximum field B_m is large compared to the time constant τ. In that case, the time derivative of the internal field inside the strand is equal to the time derivative of the external field for most of the ramp time, $\dot{B}_i \approx \dot{B}_e = B_m/T_{ramp}$. The intra-strand energy dissipation per unit volume and per ramp cycle is thus

$$Q_s = \frac{B_m^2}{\mu_0} \cdot \frac{4\tau}{T_{ramp}} . \tag{7.4}$$

This is part of the a.c. loss of the magnet.

From the equations (7.1) to (7.4) it is obvious that an untwisted superconductor (l_{twist} typically > 1000 m) would be totally useless for accelerator application.

7.2   Cable eddy currents

7.2.1   One dimensional model

Following a model originally developed by Morgan (1973) and expanded by Devred and Ogitsu (1994), we replace the Rutherford cable by a two-layer network of wires that are connected by small resistors at the cross-over points. A schematic drawing of a cable with six strands is given in Fig. 7.1. Let N be the total number of strands. Across the cable we can distinguish $N - 1$ different loops, labelled by the letter n ($1 \leq n \leq N - 1$). The magnetic flux through loop n is denoted by Φ_n. Let R_n be the resistance at the nth transition and i_n the cross-over current. Neglecting a dependence on the longitudinal coordinate we get the following set of equations for the inner loops ($2 \leq n \leq N - 2$)

$$\frac{d\Phi_n}{dt} = 2R_n i_n - R_{n-1}i_{n-1} - R_{n+1}i_{n+1} \tag{7.5}$$

while at the edges of the cable we have

$$\frac{d\Phi_1}{dt} = 2R_1 i_1 - R_2 i_2 \quad , \quad \frac{d\Phi_{N-1}}{dt} = 2R_{N-1}i_{N-1} - R_{N-2}i_{N-2} \ . \tag{7.6}$$

Let I_n be the induced current flowing in the wire section between the cross-over

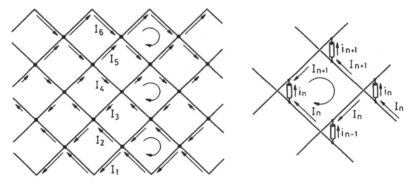

Figure 7.1: Equivalent resistive network of a Rutherford cable with $N = 6$ strands. The cross-over resistances are indicated as black dots. Also shown are the branch currents I_n along the strands as computed by Eq. (7.10). A single loop with the cross-over currents i_n is shown on the right-hand side.

points $n - 1$ and n. From Kirchhoff's law we get

$$I_n = I_{n+1} + i_n \ . \tag{7.7}$$

The sum of all currents I_n must vanish since the time-varying field is unable to induce a net current in the z direction, hence

$$\sum_{n=1}^{N} I_n = 0 \ . \tag{7.8}$$

The equations (7.5) to (7.8) lead to a set of $N - 1$ independent equations for the $N - 1$ unknowns i_n. The solutions are

$$i_1 = \frac{1}{NR_1} \sum_{m=1}^{N-1} \sum_{k=1}^{m} \frac{d\Phi_k}{dt} \quad ,$$

$$i_n = \frac{1}{R_n}\left[nR_1 i_1 - \sum_{m=1}^{n-1}\sum_{k=1}^{m}\frac{d\Phi_k}{dt}\right] \quad (2 \le n \le N-1) . \tag{7.9}$$

With the simplifying assumptions that the contact resistances at the cross-over points are all identical, $R_n = R_c$, and that moreover the magnetic flux across the cable is uniform, it is easy to express the solution in closed form

$$i_n = \frac{n}{2R_c}(N-n)\frac{d\Phi}{dt}$$

$$I_n = \frac{1}{24R_c}[(N-1)N(N+1) - 2(n-1)n(3N-2n+1)]\frac{d\Phi}{dt} . \tag{7.10}$$

Figure 7.1 shows schematically the resulting current pattern in our model cable with six strands. The induced branch currents flow in a zig-zag fashion with different directions in the lower and upper half of the cable. It is obvious that this bipolar current distribution represents a magnetic moment and will contribute to the multipole contents of the magnet. The multipole fields (see Sect. 7.2.3.) can be computed from an equivalent arrangement of linear currents.

The contact resistances lead of course to Ohmic heat generation in time-varying magnetic fields. If we call l_p the transposition pitch length of the cable, the overall heat generation per metre of cable (the a.c. loss) is given by the expression

$$G = \sum_{n=1}^{N-1} R_c i_n^2 N / l_p$$

$$= \frac{N}{4R_c l_p}\left(\frac{d\Phi}{dt}\right)^2 \frac{(N^4-1)N}{30} \tag{7.11}$$

The time derivative of the flux is

$$\frac{d\Phi}{dt} \approx \frac{w l_p}{N^2}\frac{dB}{dt}$$

where w is the cable width. For $N \gg 1$ Eq. (7.11) reduces to

$$G \approx \frac{w^2 l_p N^2}{120 R_c}\left(\frac{dB}{dt}\right)^2 . \tag{7.12}$$

7.2.2   Two dimensional model

Akhmetov, Devred and Ogitsu (1994) have extended the above analysis to the case that the inter-strand resistances and the branch currents are allowed to vary along the magnet axis. They subdivide the cable into columns along the axis which are labelled by an index k. This is sketched in Fig. 7.2.

The induced cross-over currents $i_{k,n}$ depend now on the longitudinal index k and the transverse index n. The most general case can be treated only numerically. With

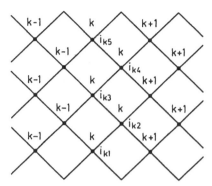

Figure 7.2: Resistive network with a longitudinal and transverse variation of the cross-over currents. Indicated are the 5 cross-over currents corresponding to the longitudinal index k.

the assumption of a uniform cross-over resistance both across and along the cable and a uniform magnetic flux along the cable, but allowing for a flux variation transverse to the cable (which is indeed the case in all magnet coils), one arrives at an interesting observation: the cross-over currents $i_{k,n}$ exhibit a longitudinal periodicity:

$$i_{k+N,n} = i_{k,n} \ .$$

The periodicity interval is identical with the cable twist pitch length l_p. We illustrate this again by studying the model cable with six strands. The equations (7.5) and (7.6) for the cross-over currents have to be modified to incorporate the longitudinal dependence.

$$
\begin{aligned}
i_{k,1} + i_{k+1,1} - i_{k,2} &= f_1 \\
i_{k,2} + i_{k+1,2} - i_{k+1,1} - i_{k+1,3} &= f_2 \\
i_{k,3} + i_{k+1,3} - i_{k,2} - i_{k,4} &= f_3 \\
i_{k,4} + i_{k+1,4} - i_{k+1,3} - i_{k+1,5} &= f_4 \\
i_{k,5} + i_{k+1,5} - i_{k,4} &= f_5 \ .
\end{aligned}
\tag{7.13}
$$

Here we have used the abbreviation $f_n = \dfrac{1}{R_c}\dfrac{d\Phi_n}{dt}$. From these equations one can compute the currents at position $k+1$ from the values at position k. For this purpose it is convenient to use matrix notation. We define a 5×5 (generally $(N-1)\times(N-1)$)

matrix $\mathbf{A}$ by

$$\mathbf{A} = \begin{bmatrix} -1 & 1 & 0 & 0 & 0 \\ -1 & 1 & -1 & 1 & 0 \\ 0 & 1 & -1 & 1 & 0 \\ 0 & 1 & -1 & 1 & -1 \\ 0 & 0 & 0 & 1 & -1 \end{bmatrix} \tag{7.14}$$

and the vectors

$$\mathbf{I}_k = \begin{pmatrix} i_{k,1} \\ i_{k,2} \\ i_{k,3} \\ i_{k,4} \\ i_{k,5} \end{pmatrix} , \quad \mathbf{F} = \begin{pmatrix} f_1 \\ f_1 + f_2 + f_3 \\ f_3 \\ f_3 + f_4 + f_5 \\ f_5 \end{pmatrix} . \tag{7.15}$$

Then the equations (7.13) read

$$\mathbf{I}_{k+1} = \mathbf{A} \cdot \mathbf{I}_k + \mathbf{F} . \tag{7.16}$$

The matrix $\mathbf{A}$ has two important properties which are easily verified by direct computation:

$$\mathbf{A}^N = 1 , \quad \sum_{k=0}^{N-1} \mathbf{A}^k = 0 .$$

Applying equation (7.16) repeatedly and using these relations we get

$$\mathbf{I}_{k+N} = \mathbf{A}^N \cdot \mathbf{I}_k + \sum_{k=0}^{N-1} \mathbf{A}^k \mathbf{F} = \mathbf{I}_k . \tag{7.17}$$

This shows explicitly that the cross-over currents are periodic in the longitudinal index k with a period N, the number of strands in the cable. This conclusion rests on the assumption that the cross-over resistances and the magnetic flux are independent of the z coordinate. In Sect. 7.2.1 the eddy currents have been taken as independent of the longitudinal coordinate z. The present treatment shows that this is a special case which will be realized only if the appropriate initial conditions are satisfied. In general the cross-over currents will vary periodically with z with a period l_p.

In addition to the cross-over resistances R_c one can also take into account the resistances R_a between adjacent strands which, however, are of minor importance concerning field distortions and a.c. losses. The most general case can only be studied in a computer simulation, see e.g. (Verweij 1995).

7.2.3 Effect of cable eddy currents on field quality and magnet performance

The small-loop cable eddy current create a bipolar current pattern as sketched in Fig. 7.1 and this will certainly generate multipole fields. In contrast to the filament magnetization currents, the eddy currents do no longer obey any coil symmetry because the cross-over resistances can vary from turn to turn and also in longitudinal direction. For this reason all allowed and unallowed multipoles are expected to appear during a ramp of the magnetic field, and their strength has to be proportional to the time derivative of B. Numerous measurements, especially from the SSC laboratory, confirm this expectation. A good example are the strongly ramp-rate dependent quadrupole and sextupole fields in an SSC dipole (Fig. 7.3). The skew multipole fields A_2, A_3, A_5, A_5 are influenced in a similar manner. The time constants for the eddy-current multipoles are in the order of a second. It is interesting to note that

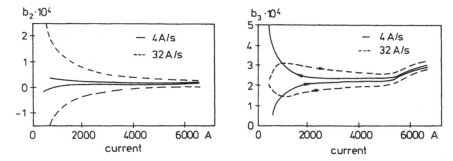

Figure 7.3: Hysteresis of the normal quadrupole and sextupole coefficients as a function of magnet current for ramp rates of 4 and 32 A/s. The ramp direction is indicated by arrows. The width of the hysteresis curves is proportional to dI/dt which proves that eddy currents are the source. The measurements were performed at Brookhaven on dipole DCA312.

the eddy-current induced sextupole is of opposite sign as the sextupole due to superconductor magnetization. The same has been observed in HERA dipoles. The apperance of 'forbidden' multipoles (normal quadrupole and skew multipoles) proves that the cable eddy currents do indeed violate the dipole coil symmetry, as stated above.

Devred and Ogitsu (1994) developed a very general computer code for the numerical calculation of the multipole fields resulting from a given arbitrary distribution of the cross-over resistances. Inversely, measured multipole fields can be used as an input to compute the cross-over resistances in all turns of the coil. These very detailed and thorough investigations were motivated by the observation of extremely large ramp-rate dependent multipoles in several SSC prototype dipoles. It was further possible to correlate the multipoles with measured a.c. losses and with a strong decrease of quench current at high ramp rates. In Fig. 7.4a the energy loss per current cycle 500 A → 5000 A → 500 A is plotted as a function of current ramp rate dI/dt. The reduction in quench current at a given ramp rate, say $dI/dt = 90$ A/s, is found to be correlated with the loss per cycle (Fig. 7.4b) which proves that eddy-current heating of the coil is responsible for the premature quenches at higher ramp rates. The magnets with a particularly strong ramp rate sensitivity were characterized by unusually low contact resistances of about 2 $\mu\Omega$.

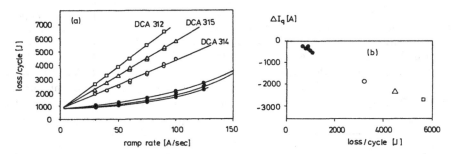

Figure 7.4: (a) Energy loss Q per cycle as a function of current ramp rate dI/dt for various SSC dipoles. Five dipoles with small losses are indicated by black dots. The three dipoles with particularly large losses are labelled by their serial numbers. (b) Correlation between quench current reduction ΔI_q and energy loss Q per cycle (both at $dI/dt = 90$ A/s). (Ozelis et al. 1993). (© 1993 IEEE)

All SSC magnets were made from cables with bare copper surface. It has been suspected that in problematic magnets like DCA312 the oxide layer formed during cable and coil production was too thin. Magnets with such an extreme ramp rate sensitivity are practically useless for an accelerator. Control of interstrand resistance is thus an important aspect of cable production. At ramp rate zero, the loss curves extrapolate to finite values which represent the hysteretic loss of the superconductor (compare Eq. (2.13)). For the current cycle chosen this loss is in the order of 600 to 800 J for the 15-m-long SSC magnets.

In Table 7.1 we reproduce part of a table from (Devred, Ogitsu 1994) showing predicted field distortions and power dissipation due to cable eddy currents in various dipole designs. The numbers refer to a uniform interstrand resistance $R_c = 10\,\mu\Omega$ and a ramp rate $dI/dt = 10$ A/s. The dipole, sextupole and decapole fields are given in 10^{-4} Tesla at a reference radius $r_0 = 25$ mm for the HERA, RHIC and Tevatron dipoles and at $r_0 = 10$ mm for the SSC dipoles (50 mm bore).

Loss measurements on a HERA dipole with ramp rates between 5 and 28 A/s and a current cycle 1500 A$\rightarrow$ 5500 A$\rightarrow$ 1500 A revealed a linear relationship between energy loss Q and ramp rate dI/dt (H. Brück and M. Stolper, private communication):

$$Q = Q_{hyst} + Q'\frac{dI}{dt} \ .$$

The constant term $Q_{hyst} = 360$ J represents the hysteretic loss of the superconductor, compare Eq. (2.13), while the slope $Q' = 7.5$ J/(A/s) describes the eddy-current loss. The instantaneous power, dissipated by the eddy currents, is thus about 90 mW in the 9-m-long magnet for a ramp rate of 10 A/s. Using this value and Table 7.1 the average cross-over resistance is found to be $R_c \approx 10\,\mu\Omega$. Then, for the standard HERA ramp rate of 10 A/s, the predicted field distortions caused by eddy currents

Table 7.1: Predicted field distortions and power dissipation due to cable eddy currents for various superconducting dipole magnet designs (Devred, Ogitsu 1994). The numbers refer to a uniform cross-over resistance $R_c = 10\,\mu\Omega$ and a ramp rate $dI/dt = 10$ A/s.

	HERA	RHIC	SSC	Tevatron
Transfer function (T/kA)	0.935	0.709	1.048	0.953
Inner cable				
width (mm)	10	9.73	12.34	7.8
twist pitch (mm)	95	74	86	57
number of strands	24	30	30	23
Outer cable				
width (mm)	10	—	11.68	7.8
twist pitch (mm)	95	—	94	57
number of strands	24	—	36	23
Field distortions				
B_1 (10^{-4} T)	1.5	0.5	6.5	0.6
B_3 (10^{-4} T)	+0.4	+0.2	+0.4	0.2
B_5 (10^{-4} T)	−0.05	−0.02	−0.02	−0.01
Power dissipation per m				
P (10^{-3} W/m)	9	3	24	4

should be in the order of $1.5 \cdot 10^{-4}$ T for the dipole and $0.4 \cdot 10^{-4}$ T for the sextupole term. Both numbers are in fairly good agreement with direct determinations of ramp-rate dependent multipoles. It is interesting to note that the hysteretic loss exceeds the eddy-current loss for the low ramp rates used in storage rings.

7.3 Eddy currents in longitudinally varying fields

7.3.1 Theoretical model

It is suggestive to search for a relation between the periodic cross-over currents in the cable and the longitudinal periodicity observed in the persistent-current multipole fields. Krempaski and Schmidt (1995) and independently Verweij and ten Kate (1995) have investigated the possibility that the time derivative of the magnetic field varies along the cable direction[1]. In accelerator magnets this is indeed realized since in the

---

[1]A similar model was presented by A.V. Tollestrup at the 1990 workshop on persistent current effects at Fermilab.

coil heads and at the current leads the magnetic field and hence dB/dt differ from the values in the straight section. Also when going from one winding turn to another the local field and its time derivative will change.

Following the lucid treatment by Krempaski and Schmidt we consider a two-strand Rutherford cable whose length l_0 is much larger than the transposition pitch length l_p. We assume that $\dot{B}$ is nonzero in a limited range of length b around the centre $z = l_0/2$ and vanishes elsewhere. The arrangement is sketched in Fig. 7.5. The

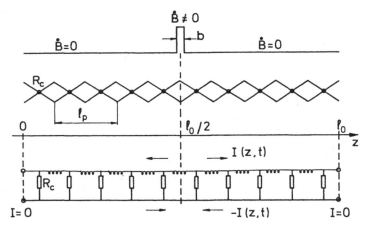

Figure 7.5: A two-strand cable subjected to a time-varying magnetic field in the centre region and the equivalent electric circuit (Krempaski and Schmidt 1995).

two wires are connected every half pitch by the contact resistance R_c. Since $l_p \ll l_0$ we may treat the two-wire system as a continuous transmission line and introduce a transverse conductance and an inductance per unit length:

$$G' = 2/(R_c\, l_p)\,, \quad L' = (\mu_0/\pi)(\ln(w/d_s) + 0.25)\,.$$

Here w is the cable width and d_s the strand diameter. The maximum induced voltage is obtained if the length b is an odd multiple of half the transposition pitch length: $U = U_{max} = l_p\, w\, \dot{B}/4$ for $b = (2k+1)l_p/2$ while $U = 0$ for $b = k \cdot l_p$.

The current $\pm I(z,t)$ in the wires must fulfill the differential equation of a transmission line.

$$\frac{\partial^2 I}{\partial z^2} = L'C'\frac{\partial^2 I}{\partial t^2} + (R'C' + L'G')\frac{\partial I}{\partial t} + R'G'I\,. \tag{7.18}$$

For vanishing longitudinal resistance and transverse capacitance ($R' = 0$, $C' = 0$) the equation reduces to

$$\frac{\partial^2 I}{\partial z^2} = L'G'\frac{\partial I}{\partial t}\,. \tag{7.19}$$

This is a diffusion-like equation with diffusivity $D = 1/(L'G')$. In the steady state, when t is large compared to the time constant of the system, the right-hand side of the equation is zero and the current is obviously a linear function of z which vanishes at $z = 0$ and $z = l_0$. The maximum $I_{max} = UG'l_0/4$ occurs around $z = l_0/2$. For a short extension of the magnetic field region, $b \ll l_0$, this is a triangle-like function[2] which can be expanded in a Fourier series of period $2l_0$.

$$I(z) = \frac{8I_{max}}{\pi^2} \sum_{n=1,3,5...} \frac{(-1)^{(n-1)/2}}{n^2} \sin\left(\frac{n\pi z}{l_0}\right) .$$

The solution $I(z,t)$ for the decay from the steady state is easy to construct. When the field increase is stopped at $t = 0$, which means that $\dot{B}$ vanishes for $t > 0$, one has to multiply the nth Fourier term with $\exp(-t/\tau_n)$. From Eq. (7.19) follows that the time constant τ_n of the Fourier term n is given by

$$\tau_n = \tau/n^2 \quad \text{with} \quad \tau = \frac{L'G' l_0^2}{\pi^2} . \tag{7.20}$$

The time constant $\tau = \tau_1$ can be quite large. If we consider a single winding turn in a 10-m-long dipole and assume a cross-over resistance $R_c = 5\,\mu\Omega$ and a transposition pitch length of $l_p = 0.1$ m then $\tau \approx 160$ s. The time constant grows with the square of the cable length. Moreover, in a multistrand Rutherford cable there are many more cross-over points than in our two-strand model, so the transverse conductance is much larger. As a consequence the time constant of a complete subcoil of the dipole may be in the order of many hours or even days.

The solution for the charging period is found in a similar way. If $\dot{B}$ is switched on at $t = 0$ the current as a function of time is

$$I(z,t) = \frac{8I_{max}}{\pi^2} \sum_{n=1,3,5...} \frac{(-1)^{(n-1)/2}}{n^2} \sin\left(\frac{n\pi z}{l_0}\right) (1 - \exp(-t/\tau_n)) . \tag{7.21}$$

If charging is stopped at $t = t_1$ the components $I_n(z,t_1)$ of the sum in Eq. (7.21) decay with their respective time constants $\tau_n = \tau/n^2$ and the current for $t > t_1$ is

$$I(z,t) = \sum_{n=1,3,5...} I_n(z,t_1) \exp(-(t - t_1)/\tau_n) . \tag{7.22}$$

Of particular interest for the accelerator is a linear ramp cycle $0 \to B_{max} \to 0$ as sketched in Fig. 7.6a. We assume a ramp-up time $t_1 = \tau/10$ and take the same interval for the dwell time at high field and the ramp-down time. Using equations (7.19) to (7.22) the current $I(z,t)$ can be computed at all times. The parameters chosen are: cable length $l_0 = 20$ m, magnetic field region $b = l_p/2 = 0.05$ m, contact resistance $R_c = 5\,\mu\Omega$, field ramp rate $\dot{B} = 0.05$ T/s. The results are plotted in Figs. 7.6b, 7.6c. When the magnetic field has undergone the ramp cycle and is kept

---

[2]The more general case of an extended region with $\dot{B} \neq 0$ is treated in (Krempaski, Schmidt 1995).

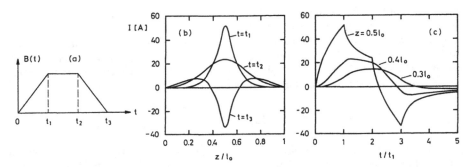

Figure 7.6: (a) Ramp cycle of magnetic field. (b) Position dependence of the current $I(z,t)$ for the times t_1, $t_2 = 2t_1$, $t_3 = 3t_1$. (c) Time dependence of $I(z,t)$ at various positions.

at $B = 0$ afterwards ($t > t_3$) there is still a considerable time dependence in the current. Eventually of course the current I approaches zero but the decay time may be many hours for a complete magnet coil. It is obvious that the zig-zag current pattern generates a magnetic field with a longitudinal periodicity.

It should be noted that the induced currents are in the 50 – 100 A range and thus not negligible in comparison with the typical transport current carried by a strand. This effect contributes to the ramp rate dependence of quench current, shown in Fig. 7.4, because the sum of transport and induced current may exceed the critical current of a strand. Eddy-current heating of the coil leads to an additional reduction of quench current.

An experiment has been performed (Verweij, ten Kate 1995) to study the effect on a 1.3-m-long Rutherford cable. The cable was clamped with 15 MPa over a length of 1.1 m to achieve a low cross-over resistance and was subjected to a time-varying magnetic field at one end. The magnetic field along the cable, measured with a Hall probe arrangement, exhibits indeed an oscillatory pattern with the cable transposition pitch as its period (see Fig. 7.7).

7.3.2 Influence on the persistent-current effects in magnets

The induction effect caused by the variation of $\dot{B}$ along the cable of a coil can qualitatively account for the longitudinal periodicity seen in all magnets. The observed multipole field pattern results from a complex superposition of the bipolar currents induced in any pair of strands and in all turns of the coil. Since the cross-over resistances and the number of pitch lengths between the coil heads will vary from turn to turn and from magnet to magnet, rather different oscillation patterns of the various multipoles must be expected for different magnets. For this reason a quantitative analysis appears not very meaningful and predictions for a new magnet will be difficult.

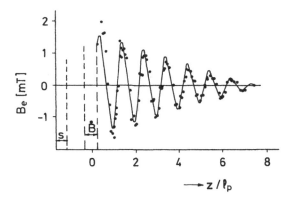

Figure 7.7: Experimental study of 'boundary-induced' eddy currents in a Rutherford cable. The time-varying magnetic field ($\dot{B} = 0.016$ T/s) is concentrated around $z = 0$. At the left end the cable has been soldered (labelled by s in the drawing) to simulate the effect of an internal solder connection in a dipole coil. The eddy-current field B_e along the cable is measured with an array of Hall probes. Continuous curve: model calculation. (Verweij, ten Kate 1995).

Nevertheless, some important features can be understood in a qualitative way. We have seen in Figs. 6.14 and 6.15 that the oscillation amplitude is very small on the initial excitation curve of a magnet but grows considerably with increasing field. This is obvious from the factor $(1 - \exp(-t/\tau_n))$ in Eq. (7.21): if the ramp time is much shorter than the characteristic time constant τ or, alternatively, if $\dot{B}$ is very small, the induced current will be tiny. On the contrary, long ramps at sufficiently large dB/dt will generate strong oscillations. Even if the field is reduced to zero afterwards or to a small value, the oscillations remain large as is evident from Fig. 7.6c.

These 'boundary-induced' eddy currents offer also a qualitative explanation why the decay rates of persistent-current multipoles depend so much on the pre-cycle (compare Sect. 6.2). If a large field sweep has been performed, significant eddy currents will be flowing in the strands whose decay times are in the order of many hours. The total current I_s in a strand is the sum of the transport current I_t and the respective eddy current I_e; it may decrease or increase in the course of time depending on the sign of I_e. What is the effect of this time dependence on the superconductor magnetization? We make the important observation that *any change* of the strand current, either positive or negative, is accompanied with a *reduction* of the overall strand magnetization. This can be understood as follows. A current change $+\Delta I_s$ creates an azimuthal magnetic field change ΔB_ϕ inside the strand. From the large previous field sweep all filaments in the strand have been left fully magnetized. In one hemisphere of the strand, ΔB_ϕ has a component parallel to the field that magnetized the strand. The filaments in this region keep their magnetization because they are already saturated. In the other hemisphere, however, ΔB_ϕ opposes the previous magnetizing field and here the filament magnetization is reduced.

For a negative current change $-\Delta I_s$ the two hemispheres are interchanged but the overall strand magnetization is reduced as well. Our conclusion is that the strong magnetization decay, observed after large field sweeps, is caused by a current redistribution among the strands in the cable. The above model provides the underlying physical mechanism. Again quantitative predictions appear difficult because of the many unknowns in the problem.

In Fig. 6.9 we have seen that there is a large magnet-to-magnet variation in the decay rates. The most likely explanation are large variations in cross-over resistances and different numbers of transposition pitch lengths between the coil ends. The significantly higher decay rates in the HERA dipoles made from the Italian LMI superconductor can be understood as well. Due to dimensional tolerances the precompression during coil curing and collaring had to be chosen almost a factor of two higher than for the coils wound from Brown Bovery conductor. It is well known (see e.g. Verweij 1995) that higher pre-compression leads to smaller cross-over resistances and hence to stronger eddy current effects.

7.4 Transmission line characteristics of a long string of magnets

7.4.1 Equivalent circuit of a dipole

A superconducting dipole cannot be considered as a pure inductance, in spite of the vanishing resistance of the conductor. We have seen in Sect. 7.2 that eddy current losses play a significant role. In an equivalent circuit they correspond to a frequency-dependent real part of the impedance. Furthermore, the tight compression of the windings, provided by the collars, introduces a sizeable capacitance. Following R. Shafer (1981) we consider as a simple model a dipole equipped with an electrically conducting beam pipe and neglect for the moment eddy currents in the cable or collars. Let L_1 be the d.c. inductance of the magnet coil (measured in the limit of vanishing frequency), l_1 its length and a its average radius. The beam pipe is influenced by the time-varying field over the same length l_1. The inductance and resistance for induced eddy currents are

$$ L_2 = \mu_0 l_1/(4\pi), \quad R_2 = \frac{2l_1\rho}{\pi\,b\,t} . $$

Here b is the pipe radius, ρ its resistivity and t the wall thickness. The mutual inductance of coil and pipe is given by the expression

$$ M = \frac{b}{a}\sqrt{L_1 L_2} . $$

Connecting the coil to an a.c. voltage $U_1 \exp i\omega t$ we obtain the equations

$$U_1 = i\omega L_1 I_1 + i\omega M I_2 \qquad 0 = (R_2 + i\omega L_2)I_2 + i\omega M I_1 .$$

The impedance of the magnet coil with inserted beam pipe is

$$Z_{coil}(\omega) = \frac{U_1}{I_1} = \left(i\omega L_1 + \frac{\omega^2 M^2}{R_2 + i\omega L_2} \right) .$$

Separating real and imaginary parts this can be written as

$$Z_{coil}(\omega) = \frac{\kappa L_1 \tau \omega^2}{1 + \omega^2 \tau^2} + i\omega \left[(1 - \kappa)L_1 + \frac{\kappa L_1}{1 + \omega^2 \tau^2} \right] . \qquad (7.23)$$

Here $\kappa = b^2/a^2$ and $\tau = L_2/R_2 = \mu_0 bt/(2\rho)$. The impedance of Eq. (7.23) corresponds to the circuit drawn in Fig. 7.8a.

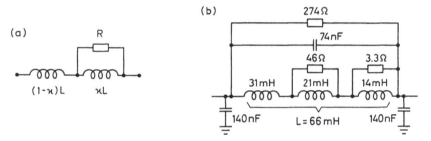

Figure 7.8: (a) Equivalent circuit for a dipole with inductively coupled conducting beam pipe. (b) Equivalent circuit of SSC dipole including cable capacitance (Smedley and Shafer 1992).

The capacity between windings can be replaced by an equivalent capacitance against ground potential. If all eddy currents are taken into consideration one arrives at the equivalent circuit in Fig. 7.8b. The frequency response of a superconducting dipole is indeed that of a damped resonance circuit, see Fig. 7.9. The parameters of the equivalent circuit have been adjusted to yield the best agreement with the measurement.

7.4.2 String of magnets

A long string of dipoles can be considered as a transmission line. In the low-frequency limit, $\omega \ll 1/\tau$, the string has a characteristic impedance and a phase velocity

$$Z = \sqrt{\frac{L'}{C'}} \approx 900 \,\Omega , \qquad v = \frac{1}{\sqrt{L'C'}}$$

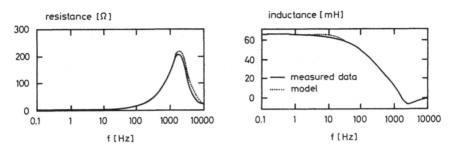

Figure 7.9: The measured resistance (real part of the impedance) and inductance (imaginary part divided by ω) of an SSC dipole as a function of frequency. The prediction based on the equivalent circuit of Fig. 7.8b is shown for comparison (Smedley and Shafer 1992).

where L' and C' are the inductance and capacitance per unit length. The phase velocity is in the order of 100 km/s or about 10000 to 20000 magnets per second.

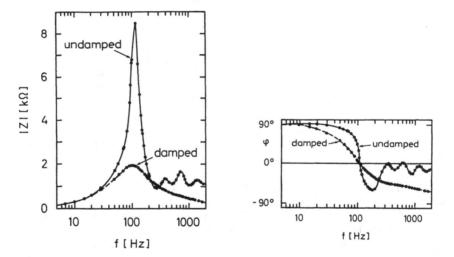

Figure 7.10: Observed resonance behaviour in a 96-magnet string in the Tevatron. Shown is the absolute value of the impedance and the phase as a function of frequency for the undamped and the damped case. Curves: model calculation. Dots: measured data. (Shafer, Smedley 1992)

At frequencies $\omega > 1/\tau$ transmission line effects become significant. Figure 7.10 shows data from a section of the Tevatron comprising 96 magnets. The absolute magnitude of the impedance and the phase are plotted as a function of frequency.

A pronounced resonance is observed at 100 Hz and some higher resonances in addition. Measurement and model calculation agree very well. Such a resonance is very undesirable as it may lead to voltage enhancements during a magnetic field ramp or to a variation of the dipole field along the chain if power supply ripple excites the resonance frequency. Fortunately it can be strongly damped by connecting the two safety current leads of a quench protection unit (compare Sect. 8.5.2) with an external resistor of about 100 Ω. The transmission-line properties of the superconducting magnet chain are a potential danger in the case of magnet quenches. Large voltage transients may propagate along the magnet chain leading to enhanced coil-to-ground voltages. An example will be presented in Sect. 8.5.

References

A. Akhmetov, A. Devred and T. Ogitsu, *Periodicity of crossover currents in a Rutherford-type cable subjected to a time-dependent magnetic field*, J. Appl. Phys. **75** (1994) 3176

A. Devred, T. Ogitsu, *Ramp rate sensitivity of SSC dipole magnets prototypes*, KEK preprint 94-156 (1994), submitted to Particle Accelerators

L. Krempaski and C. Schmidt, *Influence of a longitudinal variation of dB/dt on the magnetic field distribution of superconducting accelerator magnets*, Appl. Phys. Lett. **66** (1995) 1545, and *Theory of "supercurrents" and their influence on field quality and stability of superconducting magnets*, J. Appl. Phys. **78**(1995) 5800

G.H. Morgan, *Eddy currents in flat metal-filled superconducting braids*, J. Appl. Phys. **44** (1973) 3319

G.B.J. Mulder and E.M.J. Niessen, *Coupling losses of multifilamentary superconductors having several concentric regions and mixed matrix*, IEEE Trans. **ASC-3** (1993) 142

J.P. Ozelis et al., *A.C. loss measurements of model anf full size 50 mm SSC collider dipole magnets at Fermilab*, IEEE Trans. **ASC-3** (1993) 678

R. Shafer, *Eddy currents, dispersion relations and transient effects in superconducting magnets*, IEEE Trans. **MAG-17**(1981) 722

R. Shafer, K.M. Smedley, *Electrical characteristics of long strings of SSC superconducting dipoles*, Proc. Int. Conf. on High Energy Accelerators, Hamburg 1992, World Scientific 1993, p. 298

K.M. Smedley, R. Shafer, *Measurement of a.c. electrical characteristics of SSC superconducting dipole magnets*, Proc Int. Conf. on High Energy Accel., Hamburg 1992, World Scientific 1993, p. 629

A.P. Verweij, H.H.J. ten Kate, *Super coupling currents in Rutherford type cables due to longitudinal non-homogeneities of dB/dt*, IEEE Trans. **ASC-5** (1995) 404

A.P. Verweij, *Electrodynamics of superconducting cables in accelerator magnets*, PhD thesis, Twente University 1995, to be published

M.N. Wilson, *Superconducting Magnets*, Chap. 8, Clarendon Press, Oxford 1983

Further reading:

R. Bacher, K.-H. Mess and M. Seidel, *Transmission line characteristics of the s.c. HERA dipole and quadrupole string*, Proc Int. Conf. on High Energy Acc., Hamburg 1992, World Scientific 1993, p. 301

Chapter 8

Quenches and Magnet Protection

8.1 Transition to normal conductivity

In the discussion of quenches we confine ourselves to cables made from NbTi multifilamentary strands with a copper matrix. The critical surface of NbTi in a (T, B, J) coordinate system has been shown in Chap. 2 (Fig. 2.11). On this surface any of the three critical parameters – critical temperature, field and current density – is a function of the two other variables: $T_c = T_c(B, J)$, $B_c = B_c(T, J)$, $J_c = J_c(T, B)$. Several useful parametrizations are presented in Appendix D.

One of the first steps in the design of a superconducting accelerator is the choice of the cooling scheme and the liquid helium temperature. The accelerators Tevatron, HERA and RHIC are cooled with pressurized single-phase (supercritical) helium[1] of 4.3–4.5 K while the LHC magnet cooling is provided by superfluid helium of 1.9 K at 1 bar. Once the helium temperature has been fixed, the design values of field and current density are chosen with a sufficient safety margin below the critical surface. If by some disturbance part of a magnet coil is heated beyond the critical temperature the cable becomes normal-conducting in this region. Depending on the size of the normal zone, the available cooling may be sufficient to recover superconductivity or else the onset of Ohmic heating is so violent that the transition is irreversible and then the magnet quenches.

Basically one can distinguish two types of quenches. A 'natural' quench occurs when the nominal working point is moved across the critical surface. In magnet tests, for example, natural quenches are initiated on purpose by raising current and field simultaneously until the critical values are exceeded. Any magnet must quench under these conditions from the very definition of the critical parameters. The other type, a 'disturbance' quench, may happen with the nominal working point below the critical surface[2]. The origin of such a quench is usually a local overheating beyond the critical temperature. From Fig. 2.11 it is obvious that T_c is not a constant but depends strongly on magnetic field B and current density J. At typical operating

---

[1] A summary of liquid helium properties can be found in Appendix E.

[2] A. Devred (1992) uses the terms 'conductor-limited' and 'heat-deposited' quenches.

conditions, $T_c(B, J)$ is significantly lower than the value of 9.4 K quoted in Table 2.1 and is indeed not much above the boiling temperature of liquid helium, for instance $T_c(B, J) \approx 5$ K at $B = 5$ T and $J = 2 \cdot 10^3$ A/mm^2. The heat capacity of metals is extremely small in the 4 K regime. An energy input of a few milli-Joules per cm^3 is sufficient to raise the temperature of the conductor beyond T_c if no helium cooling is present (adiabatic case). This tiny energy corresponds to the work done by the Lorentz force if the conductor moves by just a few μm. These numbers illustrate how small the safety margin against quenches is. Conductor motion under the action of the huge magnetic forces can only be prevented if the coil is manufactured with extreme precision and clamped with a large pre-stress, see Chap. 5.

To study the properties of a composite NbTi-Cu conductor in the vicinity of the critical surface we consider a multifilamentary wire in thermal equilibrium with a helium bath ($T = T_0$). An external magnetic field B may be present but we keep it constant and suppress the B dependence of T_c and J_c in our formulae. Call A the cross-sectional area of the wire and η the volumetric proportion of the superconductor. Superconductivity prevails when the current I in the wire is lower than the critical current $I_c = A\eta J_c(T_0, B_0)$.

Now we want to introduce the important concept of *current sharing*. We have seen in Chap. 2 that close to J_c a hard superconductor exhibits a gradual transition from vanishing to high resistivity. When the current I in the composite wire is slowly increased to a value just above the critical current I_c, the superconductor enters the resistive state and transfers part of the current to the copper matrix. Because of the steeply rising resistivity (see Eq. (2.17)) the current in the NbTi is just slightly larger than I_c; in fact it is a very good approximation to assume that the current density in the NbTi stays always at the critical value and that the excess current $I - I_c$ is carried by the copper. An important observation is that both the NbTi filaments and the copper matrix are resistive in the current sharing regime and, being parallel conductors, experience the same voltage drop.

We want to analyse another way to enter the current-sharing regime which is of importance for the stability of a conductor. Let a current I flow through a composite conductor whose initial temperature is $T = T_0$. The current is confined to the NbTi and there is no power dissipation provided the critical current $I_c(T_0) = J_c(T_0) A\eta$ is not exceeded. Suppose now that some disturbance raises the temperature to such a value T_1 that the critical current drops below the applied one, $I_c(T_1) < I$. In that case the excess current has to be taken over by the copper matrix. The temperature T_{cs} at which current sharing, coupled with heat generation, sets in shall be computed as a function of the applied current I. For simplicity we assume a linear dependence of I_c on temperature (compare Appendix D)

$$I_c(T) \simeq I_c(T_0) \cdot \frac{T_c - T}{T_c - T_0} \tag{8.1}$$

and obtain

$$T_{cs}(I) = T_0 + (T_c - T_0) \cdot \left(1 - \frac{I}{I_c(T_0)}\right) . \tag{8.2}$$

For $I \to I_c(T_0)$ we get of course $T_{cs} \to T_0$.

Concerning Ohmic heating we can distinguish three temperature regimes.

(1) Below $T_{cs}(I)$ the superconductor can carry the whole current and there is no heat generation.

(2) Above T_c superconductivity vanishes and practically all current is transported by the copper matrix because the resistivity of normal NbTi is quite high. The generated power in a wire section of length l is

$$G(T) = \rho_m(T) \frac{l}{A_m} I^2 = \rho_m(T) \frac{l}{A(1-\eta)} I^2 \;.$$

The resistivity ρ_m of the copper matrix is constant up to temperatures of about 30 K.

(3) Finally, in the intermediate range $T_{cs} < T < T_c$ we have current sharing with power generation both in the NbTi and in the matrix. The voltage drop is the same in the two parallel conductors. For the matrix material it is easily calculated:

$$V(T) = \rho_m(T) \frac{l}{A_m} I_m(T) = \rho_m(T) \frac{l}{A(1-\eta)} (I - I_c(T))$$

The generated power is simply the product of voltage drop and total current

$$G(T) = V(T) \cdot I = \rho_m(T) \frac{l}{A(1-\eta)} (I - I_c(T)) \cdot I \;.$$

From the definition (8.2) of the current-sharing temperature T_{cs} it is clear that the applied current I is identical with the critical current at $T = T_{cs}$. Using again a linear dependence of critical current on temperature we get $I - I_c(T) = I_c(T_{cs}) - I_c(T) = I(T - T_{cs})/(T_c - T_{cs})$ and arrive at the following expression for the generated power

$$G(I,T) = \rho_m(T)I^2 \frac{l}{A(1-\eta)} \cdot \frac{T - T_{cs}}{T_c - T_{cs}} \;. \tag{8.3}$$

For future application it is useful to compute the power density $g = G/(Al)$ as a function of the average current density $J = I/A$ in the composite wire

$$g(J,T) = \begin{cases} 0 & \text{for} \quad T \le T_{cs} \\[2mm] \rho_m(T)J^2 \dfrac{l}{(1-\eta)} \cdot \dfrac{T - T_{cs}}{T_c - T_{cs}} & \text{for} \quad T_{cs} < T \le T_c \\[2mm] \rho_m(T)J^2 \dfrac{l}{(1-\eta)} & \text{for} \quad T > T_c. \end{cases} \tag{8.4}$$

8.2   Stability

The superconducting wires are exposed to a variety of disturbances which may heat the coil locally beyond the critical temperature: wire motion during excitation of the magnet, cracking of epoxy joints or beam losses. The stability of a coil is a measure of its ability to recover superconductivity after such a disturbance. We want to study two limiting cases:

- each wire is surrounded with liquid helium so that produced heat is directly transferred to the coolant;

- the wires are embedded in an insulating medium like epoxy, cooling of a hot spot is only possible via heat conduction along the wire.

In practice one has usually a combination of these mechanisms.

8.2.1   Cooling by heat transfer to helium

An important limiting case is that of a multifilamentary conductor immersed in a liquid helium bath of temperature T_0. Heat transfer to the helium plays the dominant role while heat conduction along the wire can often be neglected.

The principle of current sharing has been discussed in the previous section. Now we want to apply it to the special case of a wire carrying a current slightly larger than the critical current at the helium bath temperature T_0: $I \geq I_c(T_0) = J_c(T_0)A\eta$. In this case the current-sharing temperature T_{cs} coincides with T_0 and the generated power can be written as

$$G(T) = \rho_m(T)J_c^2 \frac{\eta^2}{(1-\eta)} \cdot \frac{T - T_0}{T_c - T_0} \cdot A \cdot l \, .$$

Here $J_c = J_c(T_0)$. Heat transfer to the helium bath leads to an energy flow

$$Q(T) = h\, p\, l(T - T_0)$$

where h is the heat transfer coefficient (see Appendix E) and p the 'wetted perimeter' of the wire. Superconductivity is recovered if the power generated is less than the power removed, $G < Q$. The ratio of the two quantities is called the *Stekly parameter* (Stekley and Zar 1965):

$$\alpha_{St} = \frac{G(T)}{Q(T)} = \frac{\eta^2 J_c^2 \rho A}{(1-\eta)h\, p\,(T_c - T_0)} \quad . \tag{8.5}$$

This formula applies for bath cooling. Complete cryogenic stability is achieved for $\alpha_{St} < 1$. A coil complying with this criterion can basically not quench since the copper alone can carry the current with the available helium cooling. For a strand with NbTi filaments in a copper matrix, cryogenic stability would require a copper-to-superconductor ratio of 10 or more. The large superconducting solenoid coils in storage ring experiments are often built as cryostable magnets by cladding the Rutherford cable with high purity aluminium which features a thousandfold increase in conductivity when cooled to 4.2 K and a small magneto-resistance. Cryostable or nearly cryostable magnets are discussed in greater detail in (Wilson 1983, Dresner 1995).

In accelerator magnets α_{St} is usually much larger than unity (e.g. $\alpha_{St} = 22$ in the HERA dipoles) since otherwise the coils would become extremely bulky and expensive. So these coils are definitely not cryostable and one has to accept that the dipoles and quadrupoles of a superconducting accelerator may quench, for example if energy is deposited in the coils as a consequence of beam loss. The implications for the protection system will be considered in Sect. 8.5.

Even if cooling is insufficient for excluding quenches in such a partly stabilized conductor, the helium inside a Rutherford cable has still a very beneficial effect on the safety. Baynham et al. (1981) have initiated quenches in superconducting wires by inductive heating. Figure 8.1 shows the energy density required to trigger a quench as a function of operating current. For a conductor in vacuum (adiabatic case)

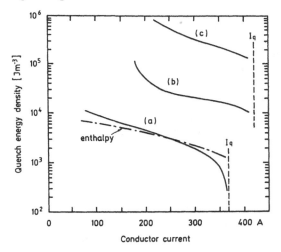

Figure 8.1: Measured energy density needed to quench a composite superconductor.
(a) Wire in vacuum. The quench energy computed from the enthalpy of the wire is indicated by the dash-dotted curve. (b) Wire surrounded by liquid helium. (c) Wire embedded in helium-filled sinter material to increase the cooled surface (Baynham, Edwards and Wilson 1981). (© 1981 IEEE)

an energy deposition of a few mJ/cm³ is sufficient to quench the conductor. With helium surrounding the wire one gains almost an order of magnitude which illustrates very clearly how important a direct contact between cable and helium is for a high performance magnet. A further improvement can be achieved by increasing the cooled surface with a helium-filled sinter material which, however, would be impractical in accelerator magnets. For very short duration of the heat pulse the helium cooling was found to be less effective.

The Rutherford-type cable with helium-transparent Kapton and glass tape insulation (see Fig. 3.9) provides optimum cooling as each wire is surrounded by liquid helium. Moreover, the enclosed helium increases the overall heat capacity of the Rutherford cable by nearly two orders of magnitude. Filling the cable with solder, which has been done in some cases to improve mechanical stability, cannot be recommended for these reasons. The same detrimental effect can be expected from the epoxy impregnation needed in Nb_3Sn coils[3].

8.2.2   Cooling by heat conduction

Minimum propagating zone

In this section we investigate superconducting wires or cables with poor contact to the helium coolant. To a certain extent this applies for fully epoxy-impregnated coils. We start with a pure NbTi wire without copper matrix. Suppose the current density in the wire is close to J_c and a section of length l has been heated by some disturbance from the helium temperature T_0 to a value above T_c. In the now normal-conducting section heat is generated which can only be removed by heat conduction along the wire. The normal zone expands if the heat generated exceeds the heat removed:

$$\rho J^2 A\, l \geq 2\lambda(T_c - T_0)A/l \ .$$

Here λ is the heat conductivity. This inequality sets a lower limit for the length the normal zone must have for developing a quench. The length of the so called *minimum propagating zone* is

$$l_{mpz} = \sqrt{\frac{2\lambda(T_c - T_0)}{\rho J^2}} \ . \tag{8.6}$$

For pure NbTi, l_{mpz} is less than 1 μm which means the normal zone will expand and the wire will quench if a section of more than 1 μm has been driven to the normal state. Obviously, pure superconducting wires are totally useless for magnets. Moreover, NbTi wires of more than 50 – 100 μm diameter are vulnerable to flux jumping (see Sect. 2.4.5).

The situation is considerably improved by embedding fine NbTi filaments in a matrix of high purity copper whose electrical and thermal conductivities at 4 K are more than thousandfold better than those of normal conducting NbTi. The minimum propagation length of a multifilamentary strand is at least a factor of thousand larger than for a pure NbTi wire.

---

[3]A recent LHC prototype dipole built in Holland reached the critical current of the cable without training but exhibited a strong ramp rate dependence, see Sect. 8.6.

Stability criterion

Now the more general case of three-dimensional heat conduction in a coil shall be addressed. Suppose a section of the coil has been raised in temperature such that Ohmic heating takes place. The generated power may be partly removed by heat conduction along the wire or by heat transfer to the coolant but usually some excess power remains which leads to a further temperature rise of the wire. The energy balance is described by the heat equation

$$C\frac{\partial T}{\partial t} = \lambda_r \left(\frac{\partial^2 T}{\partial x^2} + \frac{\partial^2 T}{\partial y^2}\right) + \lambda_z \frac{\partial^2 T}{\partial z^2} + g(J,T) - q(T) . \tag{8.7}$$

Here C is the average heat capacity of the composite conductor plus insulation, λ_z the heat conductivity in longitudinal (z) direction, λ_r the heat conductivity in radial (x, y) direction, J the current density (averaged over NbTi and copper). The term $g(J,T)$ represents the Ohmic heat generation and $q(T)$ the cooling power of the helium bath. The longitudinal heat conductivity of the composite is dominated by the copper contribution since NbTi in the normal state is a poor thermal conductor. In radial direction the heat has to cross the gaps between the wires which are filled with insulation or helium, hence $\lambda_r \ll \lambda_z$. The term $q(T)$ will be omitted as we are interested in the limiting case of negligible helium cooling.

Following Wilson (1983) we write Eq. (8.7) in a symmetric form by scaling the transverse coordinates

$$X = x\sqrt{\lambda_z/\lambda_r}, \quad Y = y\sqrt{\lambda_z/\lambda_r}, \quad Z = z .$$

Assuming furthermore that the solution is spherically symmetric in the new coordinates, the heat equation takes the simple form

$$C\frac{\partial T}{\partial t} = \lambda \frac{1}{R} \cdot \frac{\partial^2}{\partial R^2}(R \cdot T) + \rho_m J^2 \cdot \frac{1}{1-\eta} \cdot \frac{T - T_{cs}}{T_c - T_{cs}} \tag{8.8}$$

where $R = \sqrt{X^2 + Y^2 + Z^2}$ and $\lambda = \lambda_z$. The heating term has been replaced by the expression (8.4) assuming that the warm zone is in the current-sharing regime. In the scaled coordinate system the warm zone is a sphere whose origin, the point of highest temperature, can be chosen at $R = 0$. Ohmic heat is produced for all radii $R < R_1$ where R_1 is defined by the condition $T(R_1) = T_{cs}$ (compare Fig. 8.2 in the next section). This leads to the boundary condition

$$T = T_{cs} \quad \text{for } R = R_1 . \tag{8.9}$$

For $0 \le R \le R_1$, the solution of Eq. (8.8) which remains finite for $R \to 0$ and fulfils the boundary condition $T = T_{cs}$ at $R = R_1$ has the following form

$$T(R,t) = T_{cs} + \frac{c_1}{R}\sin\left(\frac{\pi}{R_1}R\right)\exp(\beta t) . \tag{8.10}$$

Inserting into (8.8) we get the following relation for the yet unknown parameter β

$$C\beta = -\lambda(\pi/R_1)^2 - \frac{\rho_m J^2}{(1-\eta)(T_c - T_{cs})} \; . \qquad (8.11)$$

From the equations (8.10) and (8.11) we derive a stability criterion:
the parameter β must be negative since obviously for $\beta < 0$ the temperature excursion
(8.10) vanishes with time while for $\beta > 0$ the temperature would rise exponentially.
So the limit of stability is given by the condition $\beta = 0$. This defines the minimum
diameter $2R_1$ the heated zone must have for developing an instability. Per definition
this is identical with the minimum propagating zone

$$2R_1 \geq l_{mpz} = \pi \left(\frac{\lambda(T_c - T_0)}{\rho J^2} \right)^{1/2} \; . \qquad (8.12)$$

This is almost the same result as in the one-dimensional case, compare Eq. (8.6).
The minimum propagating zone is a sphere in the scaled coordinates and a rotational
ellipsoid in the real coil with a large half axis $a = R_1$ along the wires and a small half
axis $b = R_1\sqrt{\lambda_r/\lambda_z}$ in transverse direction.

Here it must be emphasized that our treatment of the time-dependent heat equation is not suited for determining the time evolution of the warm zone. The reason
is that the limiting radius R_1 of the heat-generating zone will move as a function
of time – either decrease for a vanishing disturbance or grow for a disturbance that
develops into a quench. No analytical solution is available for the heat equation with
time-dependent boundary conditions. Moreover, Eq. (8.8) applies only for $T < T_c$,
and this condition is valid only at the very start of a quench. For a proper description
of a propagating quench front, temperature-dependent heating and cooling conditions
must be taken into account.

What is then the value of the above treatment? It has the merit of showing very
clearly the *onset of instability* and setting a criterion for this. The observation that
this can be done analytically seems quite satisfying to us[4]. Incidentally, the investigation of collective instabilities in high-intensity particle beams follows a similar line
of argumentation.

Threshold energy for quenching

The stationary case $\beta = 0$ shall be investigated in more detail. The function

$$T(R) = T_{cs} + \frac{c_1}{R} \sin\left(\frac{\pi}{R_1} R \right)$$

is the solution of the time-independent heat equation for $0 \leq R \leq R_1$. In the region
$R > R_1$ the last term in the differential equation (8.8) is missing because no heat is

---

[4]L. Dresner (1995) proves that the conditions leading to a quench are not identical to those for
quench recovery. Unfortunately, the treatment is far from being elementary.

generated when $T < T_{cs}$. The solution is here

$$T(R) = c_2/R + c_3 \quad \text{for } R > R_1 .\tag{8.13}$$

To determine the unknown constants c_1, c_2, c_3 one has to impose continuity on $T(R)$ and $\partial T/\partial R$ at $R = R_1$ and the boundary condition $T(R) \to T_0$ for large R. Mathematically one would take the limit $R \to \infty$ but in practice the warm zone extends for at most a few centimetres since then the helium bath is reached. At this point the temperature gradient may have a discontinuity.

The 'cold' boundary determines the constant c_1 and hence the maximum temperature inside the warm zone. If $T(R)$ reaches the bath temperature T_0 at $R_2 = \xi R_1$ with a yet unknown parameter $\xi > 1$, the temperature profile is computed to be

$$T(R) = \begin{cases} T_{cs} + \dfrac{\xi(T_{cs} - T_0)R_1}{(\xi - 1)\pi R} \sin\left(\dfrac{\pi}{R_1}R\right) & \text{for } R \le R_1 \\[2ex] T_{cs} - \dfrac{\xi(T_{cs} - T_0)(R - R_1)}{(\xi - 1)R} & \text{for } R_1 < R \le R_2 \\[2ex] T_0 & \text{for } R \ge R_2 \ . \end{cases}\tag{8.14}$$

Figure 8.2 shows as an example the calculated temperature profiles for $\xi = 1.25$, 1.5, 1.75, 2.0 in an SSC dipole at $T_0 = 4.35$ K and $B = 6.5$ T. The most likely profile corresponds to $\xi \approx 1.5$ as explained below.

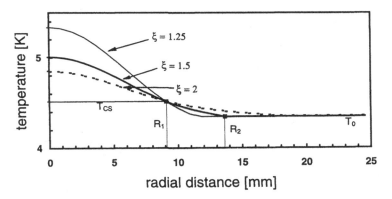

Figure 8.2: The computed temperature distribution as a function of radius for various values of the parameter ξ, see text.

The energy needed to create a minimum propagating zone depends on the actual temperature profile and hence on the point $R_2 = \xi R_1$ where the bath temperature is reached. It is given by the double integral

$$E_{mpz} = \int_0^{R_2} \left(\int_{T_0}^{T(R)} C(T')dT' \right) 4\pi(\lambda_r/\lambda_z)R^2 dR .\tag{8.15}$$

The factor (λ_r/λ_z) rescales the transverse coordinates X, Y to the physical coordinates x, y. The integral cannot be evaluated analytically. Numerical integration shows that E_{mpz} as a function of the parameter ξ has a shallow minimum at $\xi \approx 1.5$ with a variation of less than 20% in the range $1.4 < \xi < 2.5$.

In the absence of direct helium cooling, E_{mpz} is the threshold energy for initiating a quench. With helium cooling present, as is the case in coils wound from Rutherford cable, the trigger threshold may be considerably higher (compare Fig. 8.1) but even then this quantity is a good measure for the relative stability of various coil and cable configurations or different operating points. The computed energies E_{mpz} for the HERA and SSC dipoles are plotted in Fig. 8.3a as a function of magnetic field. A rapid decrease is observed when the maximum field is approached. The SSC magnet reaches higher fields owing to the small inner bore of 40 mm but in addition the design field is much closer to the maximum field. As a consequence the threshold energy for creating a minimum propagating zone drops to the tiny value of 10 μJ in agreement with other calculations (Ng 1988). The minimum propagating zone at a local field of 6.9 T is smaller than a strand diameter hence helium cooling is rather ineffective. For this reason our adiabatic calculation should yield a good estimate of the energy needed to trigger a quench. The HERA design field of 4.7 T (with a local field maximum of 5 T in the coil) is more conservative and corresponds to an energy of 150 μJ for creating a minimum propagating zone. This zone extends over some 10 mm, so helium cooling can be expected to raise the quench-threshold energy above E_{mpz}. A moderate improvement of stability is achieved by lowering the bath temperature, see Fig. 8.3b, but a reduction in nominal field is far more effective.

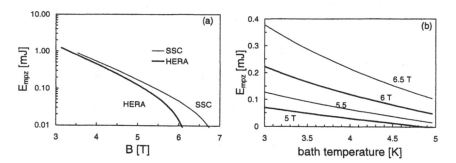

Figure 8.3: (a) Computed energy deposition E_{mpz} needed to create a minimum propagating zone as a function of local magnetic field in the coil of a HERA and an SSC dipole. (b) Dependence of E_{mpz} on helium bath temperature.

The cooling of an accelerator dipole is based on both mechanisms, heat conduction and heat transfer to helium. For short heat pulses, for instance from sudden beam losses, the adiabatic heating dominates. The helium contained in the voids of the Rutherford cable plays a role, however, as it enhances the overall heat capacity of the cable by nearly two orders of magnitude.

8.3   Quench propagation

In an accelerator the magnets are connected in series and when one magnet in the chain quenches it has to absorb most of its own field energy which may be in the Mega-Joule range. A rapid propagation of the normal zone is of great importance. The huge magnetic energy must be distributed over a sizeable fraction of the coil volume to prevent local overheating and possible destruction of the conductor. The quench propagation velocity is therefore an important property of a superconducting coil[5].

Concerning the quench properties of accelerator magnets there are in fact two conflicting requirements: the threshold energy for iniating a quench should certainly be as high as possible but once a quench has happened it should be rapidly spread over the whole coil. The first requirement calls for a large copper-to-superconductor ratio while the propagation of the warm zone is fastest with a small copper proportion in the composite. Once a sufficiently long normal zone has been created it will grow until the whole magnet has quenched or until current density and magnetic field have dropped below a level where Ohmic heating is surpassed by cooling.

8.3.1   Calculation of quench propagation velocity

Adiabatic case

In cable samples the normal zone is found to expand with constant velocity, except in the immediate vicinity of the quench origin. This simple behaviour will be taken as basis of a model description. The quench propagates mainly along the cable; transverse propagation is impeded by the insulation and the helium contents in the cable. To derive an expression for the velocity in the adiabatic case, we start from the time-dependent heat equation (8.7) but disregard the transverse coordinates and the helium cooling term:

$$C\frac{\partial T}{\partial t} = \lambda\frac{\partial^2 T}{\partial z^2} + g(J,T)\,.$$

To find a solution the assumption is made that the warm front moves with a constant velocity v along the positive z direction. Since expansion takes also place in the opposite direction the length of the warm zone expands with $2v$. It is convenient to rewrite the differential equation in a co-moving coordinate system:

$$\zeta = z - z_w = z - vt$$

---

[5]The 14-m-long LHC dipoles contain a stored energy of 7 MJ at the design field of 8.4 T. Several magnets are grouped together in a protection unit. Natural quench propagation would be too slow to prevent burnout of the coil in case of a localized quench. For this reason all dipoles in a unit will be quenched along their entire length by firing heaters.

where $z_w = vt$ is the position of the warm front as a function of time. Then the partial differential equation reduces to an ordinary differential equation

$$\lambda \frac{d^2 T}{d\zeta^2} + Cv \frac{dT}{d\zeta} + g(\zeta) = 0 \ . \tag{8.16}$$

Ignoring current sharing, the heat generation is approximated by a step function

$$g(\zeta) = \begin{cases} \rho J^2 & \text{for } \zeta < 0 \\ 0 & \text{for } \zeta > 0 \end{cases}$$

and the temperature at the centre of the warm front is written as

$$T_w = T(\zeta = 0) = 0.5(T_c + T_0) \ .$$

Equation (8.16) is solved by

$$T(\zeta) = \begin{cases} T_c - (T_c - T_w) \exp(a\zeta) & \text{for } \zeta < 0 \\ T_0 + (T_w - T_0) \exp(-b\zeta) & \text{for } \zeta > 0 \ . \end{cases} \tag{8.17}$$

The continuity of $T(\zeta)$ at $\zeta = 0$ is obviously fulfilled. Requiring in addition the continuity of the temperature gradient $dT/d\zeta$ (the heat flux) and using Eq. (8.16) for determining the unknown constants a and b, the velocity v can be computed.

$$v \equiv v_{adiab} = \frac{J}{C} \sqrt{\frac{\rho\lambda}{T_c - T_0}} \ . \tag{8.18}$$

This is the quench propagation velocity in the adiabatic limit, neglecting heat transfer to the helium surrounding the strands. Moreover the heat capacity C and the longitudinal heat conductivity λ have been treated as independent of T. Inserting reasonable numbers yields velocities in the order of 10 – 50 m/s.

Dresner (1994) assumes a T^3 dependence of the specific heat in Eq. (8.16) and obtains

$$v = v_{adiab} \sqrt{\frac{4 T_0^5 (T_c - T_0)}{T_c^2 (T_c^4 - T_0^4)}} \ . \tag{8.19}$$

Corrections due to helium cooling

When helium cooling is available the speed is expected to be somewhat lower than in the adiabatic case. M.N. Wilson (1983) derives a correction

$$v = v_{adiab} \frac{1 - 2y}{\sqrt{yz^2 + z + 1 - y}} \tag{8.20}$$

where the terms y and z arise from steady-state and transient heat transfer, respectively, and are given by

$$y = \frac{hp(T_w - T_0)}{AJ_{avg}^2 \rho_{avg}} , \quad z = \frac{Q_{lat}}{C_{avg}(T_w - T_0)} \ .$$

Here Q_{lat} is a latent heat term which characterizes the energy needed to establish a vapour film on transition from nucleate boiling to film boiling. The specific heat C has to be averaged over the temperature range of the transition and over the materials in the coil.

Although equations (8.19) and (8.20) are based on different assumptions the resulting quench velocities differ by just a few % for realistic values of T_c and T_0. For further approximate relations we refer to (Dresner 1994, 1995), (Lvovski and Lutset 1982), (Cherry and Gittelman1960), (Devred1989). In the next section we will see that the observed quench propagation velocities in long magnets are considerably larger than predicted. Quench propagation in pulsed and a.c. magnets is treated in (Mulder et al. 1992).

8.3.2 Experimental methods and results on quench propagation

Quench velocities are easily measurable in a piece of cable exposed to an external magnetic field. The quench is started by a heater pulse at one end and voltage taps along the cable are used to detect the arrival of the warm front. An example of such a measurement is shown in Fig. 8.4. Measurements in complete magnets are

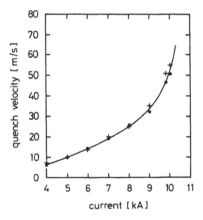

Figure 8.4: Quench propagation velocity in two cable samples, plotted as a function of current (Ghosh et al. 1989). (© 1989 IEEE)

more demanding. Several techniques have been applied. In the SSC research and development program a number of 17-m-long prototype dipoles were equipped with voltage taps along the coil. The wires were attached before assembly of the half-coils which required great care to avoid damage during coil compression in the curing and collaring press. For tests with a short HERA prototype dipole two devices with retractable needles were inserted into the bore, and by means of a key the needles

were driven out and pierced into selected turns (Bonmann et al. 1987). Both methods
are inadequate for series magnets because the danger of damaging the insulation or
producing shorts between turns is too high.

Non-destructive methods are based on the detection of ultrasonic noise or on
electromagnetic induction. A quench is a violent event which generates ultrasonic
noise. Acoustic measurements with several microphones can reveal the origin and
propagation of a quench (Nomura et al. 1980). The drawback of this method is
that such noise is produced also at normal excitation of a magnet without quench
(Chikaba et al. 1990).

By far the most elegant and versatile instrument for detecting and localizing
quenches is the so-called *quench antenna* devised by Krzywinski (Leroy et al. 1993).
It has been successfully applied at CERN (Siemko et al. 1993), the SSC laboratory
(Ogitsu et al. 1993), (Ogitsu 1994) and BNL (Ogitsu, Ganetis et al. 1995). It
is based on the idea that a quench is accompanied with current transfer from one
strand to another or some small wire motion, so in either case with a time-varying
magnetic field which should be detectable by pick-up coils inside the bore tube of the
magnet. To increase sensitivity the large signal from the decaying main dipole field is
internally compensated. This can be accomplished by using a suitable combination of
either radial or multipolar pick-up coils. With both types of antennas the azimuthal
position of a quench in a dipole magnet can be located. For this purpose one needs
two sets of two radial coils, rotated against each other by 90°, or four multipolar coils
(normal and skew quadrupole, normal and skew sextupole) which are all mounted on
a common body. At the SSC laboratory this technique has been developed to such
an accuracy that the origin of a quench could be located at a particular winding turn.

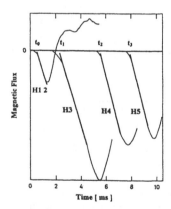

Figure 8.5: Time sequence of induced signals in four quench antennas in a 1-m-long LHC
prototype dipole (Siemko et al. 1993).

The quench velocity can be derived from the voltage rise in the antenna signal and from the time delay between the induced signals in different antennas along the magnet axis. The longitudinal accuracy is in the order of 10 mm. An example of a measurement with four antennas in a 1-m-long LHC prototype dipole is shown in Fig. 8.5. The delays are clearly observable. Each antenna signal exhibits a linear rise which corresponds to the time the warm front needs to move through the cable section subtended by the antenna.

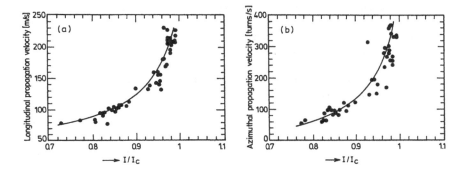

Figure 8.6: (a) Longitudinal quench propagation velocity in SSC dipoles as a function of I/I_c. (b) Azimuthal (turn-to-turn) propagation velocity as a function of I/I_c (Devred, Chapman et al. 1989a).

Data from 17-m-long SSC prototype dipoles, obtained with the voltage-tap technique, are depicted in Fig. 8.6. With increasing coil current the longitudinal quench propagation velocities rise rapidly and reach values of more than 200 m/s close to the critical current. This enormous speed is very fortunate for spreading the dissipated energy. The azimuthal (turn-to-turn) propagation is considerably slower due to the insulation in between the windings but still around 200 turns per second.

The surprisingly high quench propagation velocities near the critical current cannot be explained in terms of an adiabatic model. It has been speculated (Dresner et al. 1990) that a thermohydraulic effect in liquid helium is the underlying mechanism. Measurements at the SSC laboratory have shown that pressure waves in 4.3 K helium propagate with a speed of up to 230 m/s. Suppose the coil is operated close to J_c and a quench happens at one end. The sudden release of energy creates a pressure wave moving with a speed of about 200 m/s along the coil. Associated with the pressure rise is a temperature excursion that may exceed the critical temperature $T_c(B, J)$ at the instantaneous values of magnetic field and current density and thus induce a quenching zone co-moving with the wave front. The phenomenon is described in (Dresner 1995).

8.4   Heating of the coil after a quench

8.4.1   Experimental results

We have seen that economic and spatial reasons exclude the use of a fully cryostable conductor in accelerator magnets. It is then mandatory to investigate how much the coil may heat up after a quench. A very conservative limit is 100 K because then the thermal expansion is very small and mechanical stress in the coil and support structure is avoided. Common practice is, however, to go well beyond this point and the experience with hundreds of magnets has shown that this is permissible without risking degradation. For the HERA magnets a limit of 450 K has been set, well below the melting temperature of the solder joints.

Once a quench has been detected the power supply is switched off and the stored magnet energy is dissipated in a dump resistor and in the normal-conducting part of the coil. The current decays with a typical time constant of a few 100 ms. A relation can be established between the time dependence of the current after a quench and the highest temperature in the coil. The power density in a normal coil section is $\rho(T)J^2(t)$; during a time interval dt the section is heated by

$$dT = \frac{1}{C(T)}\rho(T)J^2(t)dt \ .$$

Note that resistivity ρ and heat capacity C depend on temperature T and have to be averaged over the materials in the coil while the current density J is a function of time. Separation of variables and integration yields

$$\int_0^\infty J^2(t)dt = \int_{T_0}^{T_{max}} \frac{C(T)}{\rho(T)}dT = F(T_{max}) \ . \tag{8.21}$$

From known material properties the integral on the right hand side can be evaluated as a function of the maximum temperature T_{max}. The integral on the left hand side can be measured. This way it is possible to establish a relation between the time integral of J^2 and the so-called *hot spot temperature* T_{max}, the highest temperature in the coil. Figure 8.7a shows experimentally determined hot spot temperatures as a function of the integral over the squared magnet current. The hot spot temperature depends nearly quadratically on this quantity. The time integral over the square of the magnet current is usually quoted in units of 10^6 A^2 s (this unit is sometimes abbreviated as 'MIIT').

The SSC magnet group has performed systematic quench studies (Devred 1989) on 17-m-long dipoles which were equipped with small heaters to initiate local quenches and with voltage taps to detect the development of resistive voltages along the windings. The temperature rise of a quenched section in the inner coil layer is plotted in Fig. 8.7b as a function of time after the quench has been detected and the power supply switched off. Within 0.3 s the temperature reaches a plateau at 150 K which

is a moderate level for a superconducting coil. The data agree well with model calculations.

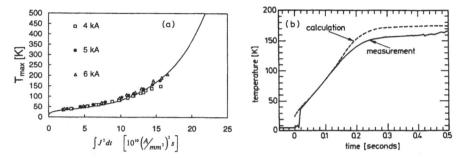

Figure 8.7: (a) Measured and calculated hot spot temperatures in a 1-m-long dipole as a
function $\int J^2 dt$ (Bonmann et al. 1987). (b) Time evolution of hot-spot temperature during
a quench induced at 6500 A on turn 1 of the inner coil of an SSC dipole (Devred, Chapman
et al. 1989b).

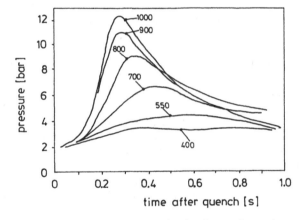

Figure 8.8: Helium pressure rise in a Tevatron dipole after a triggered quench. The numbers
at the curves refer to the corresponding Tevatron energies in GeV (Theilacker, Norris, Soyars
1994).

A consequence of coil heating is a pressure rise in the liquid helium. For an
envisaged upgrade of the Tevatron, a series of quench tests were performed on a
Tevatron dipole at field levels corresponding to proton/antiproton energies ranging
from 400 to 1000 GeV. The recorded pressure rises after a triggered quench are plotted
in Fig. 8.8 as a function of time. The maximum pressure rise of about 12 bar at

the highest excitation does not constitute a danger for the pipes and vessels of the cryostat.

Another concern are force transients in the coil and collar structure during coil heating. A calculation by Perini and Rodriguez-Mateos (1992) for the LHC dipole shows that within 250 ms the compressive stress in the inner coil layer changes by up to 40 MPa. While this number is safely below the tolerable limit in the static case, there may be fatigue effects if quenches should happen frequently.

8.4.2 Numerical methods

The heat equation (8.7) cannot be integrated analytically even with the simplifying assumption of temperature-independent material properties. Also external actions like the firing of quench heaters or the switching of bypass thyristors are unaccessible in an analytical treatment. A number of numerical codes have been written to simulate quenches on the computer. The first program published was named 'QUENCH' (Wilson 1968). Starting with a current I_0 at time t_0 the quench velocity is calculated using the material properties evaluated at the instantaneous values of magnetic field and temperature. Assuming a constant expansion speed the volume V_1 of the warm zone is calculated at the next time step t_1. The average temperature in V_1 is determined and the material properties are recalculated. A new normal conducting layer is added, and the temperatures in the inner layers are updated. Each layer keeps its own record of temperature history. Magnetic field and temperature distribution determine the total coil resistance and hence the resistive voltage drop at any time interval. A simple Euler algorithm with variable material constants converges reasonably well provided the time steps are small enough.

In a similar fashion K. Koepke (1980) predicted successfully the behaviour of the Tevatron magnets. These programs as well as modified versions of QUENCH for HERA (Otterpohl 1984) or QUENCH-M (Tominaka et al. 1992) are FORTRAN routines adapted to a particular magnet type and protection circuitry.

Pissanetzky and Latypor (1994) developed a more general code, applicable to magnets with single or multiple coils, with or without iron yoke, operating in the persistent mode or from external power supplies. The quench is allowed to start at an arbitrary point of the coil and is propagated in three dimensions. Multiple independent fronts can coexist. Local magnetic fields and inductive couplings of the coils are calculated by the finite element method.

For the LHC magnets Hagedorn and Rodriguez-Mateos (1992) designed another generally applicable simulation package called QUABER, which is based on the professional tool SABER (trademark of Analogy Inc.).

The programs mentioned above and many variants are able to describe a 'typical' quench fairly well but the results depend critically on the input assumptions. It is advisable to investigate a 'reasonable' range of input parameters to get a feeling for the uncertainty of the predictions. Measurements on full-size magnets and magnet strings are indispensable to check the validity of the numerical codes and establish the regime of safe operation of the superconducting accelerator. Such tests have been or are being performed at Fermilab, DESY, Brookhaven and CERN.

8.5 Quench detection and magnet protection

8.5.1 Quench detection and protection of a single magnet

The maximum coil temperature T_{max} after a quench (Eq. (8.21)) depends primarily on the decay time of the current. Time constants of less than 1 s are needed to limit T_{max} to values below 300 K. Hence the coil current has to be switched off in a few hundred milliseconds when a quench occurs. This requires a fast detection of the quench which is normally done by recording the resistive voltage $U_Q = R_Q I$ that builds up when a normal zone is created. The rising resistance, however, leads to a current decline and thus to a superimposed inductive voltage which has to be eliminated.

The stored magnetic energy has to be safely dissipated. For a single magnet of inductance L_1 this is easily accomplished by connecting the coil to a dump resistor R whose resistance is chosen to yield the desired decay time constant $\tau = L_1/R$ of 0.2–0.5 s typically. Much shorter decay times are dangerous since excessive internal voltages would arise even if the voltage across the coil terminals is kept close to zero by clamping diodes in the power supply. Figure 8.9 shows the equivalent electrical circuit of a quenching magnet. The growing resistance R_Q of the quenched section separates the coil inductance into two parts with a mutual coupling. Neglecting the mutual inductance, the internal voltage is

$$V_Q = I R_Q \,.$$

Internal voltages of several hundred volts are typical for high initial currents which decay within 0.3 to 1 s.

8.5.2 Protection of a string of magnets

An accelerator consists of a large number of magnets connected in series and magnet protection is a serious challenge since it is impossible to discharge the magnets across individual dump resistors. Treating the entire ring as one huge superconducting coil is also impossible as we will demonstrate for the HERA case. The inductance in the HERA ring adds up to $L=26.5$ H. At the design field of 4.7 T the stored magnetic energy amounts to 330 MJ, an amount sufficient to melt about 500 kg of copper. If one tried to extract the energy via a single external resistor its resistance would have to be around 50 Ω for a decay time constant of 0.5 s and the voltage against ground potential would exceed 250 kV.

The recipe is therefore:

- detect the quench,

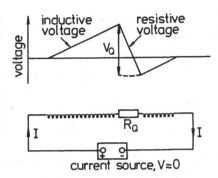

Figure 8.9: Equivalent circuit diagram of a quenching magnet and distribution of inductive and resistive voltages.

- isolate the quenching magnet,

- spread the stored magnetic energy,

- subdivide the inductance as much as technically feasible.

Practical quench detection systems

The quench detection system of the Tevatron, the first large superconducting accelerator, is based on the measurement of potential differences. Average values of the inductive voltages during ramps are calculated and used for comparison with the measured data. A significant discrepancy indicates a quench. The large resistance of the voltage taps, chosen for safety reasons, and the cable capacitance introduce a sizeable signal distortion that has to be corrected for.

The system developed for HERA uses a bridge circuit for each dipole[6]. The inductive voltages in the two half-coils compensate each other perfectly and hence the bridge is very sensitive to small resistive voltages which are in most practical cases quite different in the half-coils. Additional bridges covering groups of magnets provide redundancy. A radiation-resistant magnetic isolation amplifier with low noise sensitivity amplifies the bridge current. A similar but more versatile version has been proposed for UNK (Bolotin et al. 1992). For the LHC a bridge circuit is under discussion but the presently favoured scheme is based on semiconductor isolation amplifiers.

Current bypass at a quenched magnet

In a long string of magnets the inductivity becomes large and a long decay time (for instance $\tau = 25$ s in HERA) has to be chosen to avoid excessive inductive

---

[6]The superconducting cable connecting to the next magnet is included in the circuit.

voltages against ground potential. The current in the string has to be guided around a quenched magnet to prevent destruction of the coil. The equivalent circuit is shown in Fig. 8.10.

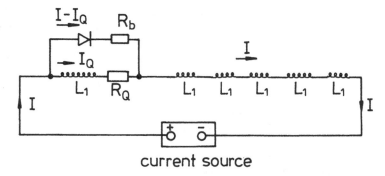

Figure 8.10: Protection of a quenching magnet (inductivity L_1) in a long string of magnets (inductivity $L = NL_1$). Most of the current is bypassed around the quenched magnet. The bypass branch may contain either a diode or a thyristor. R_b is the resistance of the bypass branch.

The total inductance L of the magnet string is much larger than the inductance L_1 of a single magnet, hence the main current I decays with a much longer time constant than the current I_Q in the quenching magnet. The differential equation for I_Q is

$$L_1 \frac{dI_Q}{dt} + I_Q R_Q(t) = (I - I_Q) R_b \ . \tag{8.22}$$

Since $R_Q(t)$ grows with time an analytic solution is not available. But once the whole coil has become normal one arrives at a steady state solution

$$I_Q = I \frac{R_b}{R_b + R_Q} \approx I \frac{R_b}{R_Q} \ . \tag{8.23}$$

To minimize I_Q the resistor R_b in the bypass line should be made as small as possible.

Two solutions have been realized. The Fermilab Tevatron has a short ramp time of 7–10 s with large inductive voltages, and thyristors act as fast switches. The protection unit comprises four dipoles and one quadrupole. The Tevatron magnets are characterized by a special feature: current input and output are at different ends of the magnet. This implies that a half-turn is missing in the coil. This half-turn is provided by the return conductor. The electrical connections alternate from half-cell to half-cell of the FODO lattice. This scheme ensures that forward and backward current decay synchronously which is important to avoid asymmetry forces between coil and iron yoke. Heaters are needed to distribute the stored energy over all magnets in a half-cell. The slowly decaying main current as well as the stored energy of the other magnets is bypassed through the thyristors. Operating only at room temperature the thyristors have to be mounted outside the cryostat and hence current feedthroughs are needed. These require a very careful design since their electrical resistance (which is the main contribution to R_b) should be minimized while the thermal resistance should be large to reduce the heat load on the liquid helium system. During a quench the current leads heat up considerably with the danger of quenches in the connecting superconducting cables[7]. A fast recooling time is another important design criterion. Current feedthroughs made from high-T_c superconductor may improve the situation considerably.

In storage rings with a long ramp time (e.g. 600 s in HERA) and correspondingly low inductive voltages, the thyristors can be replaced with diodes. Cottingham (1971) first proposed to mount silicon diodes inside the liquid helium cryostat. At 4 K the threshold voltage is around 4 V but it drops to the usual 0.7 V when current is drawn and the semiconductor warms up. This scheme offers several advantages: the bypass resistance is much smaller; there is no heat load on the cryogenic system from current feedthroughs; each magnet can be bypassed by its own diode; finally, leaving out the current feedthroughs makes the cryostats easier to build and cheaper. The cold-diode concept has been successfully adopted for HERA and is foreseen for RHIC and LHC. It is not suitable for rapidly cycling machines since the inductive voltage across a magnet would exceed by far the threshold voltage of the diode and a large part of the current would commute into the bypass branch.

The bypass diode has to be selected with great care. A low dynamic resistance, radiation hardness and a sufficiently high backward voltage are required. In the Large Hadron Collider the high magnet current of 11.5 kA and the expected radiation dose put severe constraints, but nevertheless promising solutions have been found (Hagedorn and Coull 1994). The two coils of a twin-aperture dipole are excited by the same current and will be bypassed by two thin-base epitaxial diodes connected in series. The diodes are expected to survive the estimated dose of 35 kGy and $1.5 \cdot 10^{14}$ neutrons per cm^2 for a 10-year running period.

Spreading of the stored energy

At their low design field of 3.5 T the RHIC magnets are operated with a safety margin of more than 30%. There is no danger of coil overheating in case of a localized quench and heaters to spread the normal zone are unnecessary. The same applies for the HERA proton storage ring when operated up to the nominal field of 4.7 T (safety margin 25%) but to leave room for a possible upgrade quench heaters were installed in all dipoles. The RHIC and HERA magnets are bypassed by individual diodes. In the Tevatron the thyristor branch bypasses several magnets, and a spreading of the normal zone by firing heater strips is essential for magnet safety. In the LHC, finally, the operating current and the stored energy are so enormous that an artificial quench

---

[7]Large helium-cooled copper blocks, called 'quench stoppers', are attached to the current leads to avoid heating of the superconductor.

spreading over all magnets in a protection unit is absolutely vital for preventing magnet burnout.

Quench heaters should be in close thermal contact with the coil, the optimal position would in fact be in between the two coil layers, but a compromise must be found to preserve the integrity of the electrical insulation. For the latter reason the heater strips are mounted on the outer surface of the second coil layer. The energy to fire the heaters is stored in capacitor banks which are located in the accelerator tunnel or in auxiliary buildings.

Subdivision of the magnet ring

One of the most important safety measures is a subdivision of the magnet ring into sections of lower inductance. The Tevatron consists of 24 individually powered sections. Such a favourable solution was not possible for HERA because the machine is installed in an underground tunnel without access at the centre of the 90° arcs. The HERA ring is subdivided into eight dipole sections (45° octants) and a ninth section comprising all quadrupoles. Dump resistors are installed in between any two sections. Under normal operating conditions, the dump resistors are bridged by switches which are opened when a quench is detected. The midpoints of the resistors stay on virtual ground potential. The voltage in any magnet is then restricted to an acceptable level of ±500 V against ground. To cope with the possibility that one of the switches fails to open, thereby breaking the symmetry, an equalizing line has been installed. When a quench happens the midpoints of the dump resistors are connected to the equalizing line and are thus pulled to ground potential. For the two magnet rings of RHIC a similar scheme is foreseen. Each ring is fed by its own power supply.

In Sect. 7.4 we have studied the transmission lines effects of the magnet string and mentioned the danger of excessive transient voltages if a quench happens. Figure 8.11 shows the recorded voltages to ground potential at the dump resistor between two HERA octants. In Fig. 8.11a the equalizing line was missing. Large voltage

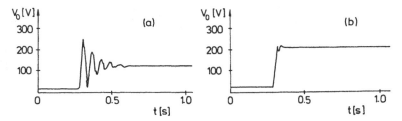

Figure 8.11: Recorded voltage V_0 to ground potential in the HERA magnet chain following a quench. (a) Without equalizing line, (b) with equalizing line. (R. Bacher, private communication)

overshoots are seen. After installation of the equalizing line the overshoots are no longer present (Fig. 8.11b).

For the LHC it has been proposed (LHC 1995) to divide the ring into 8 independent units (each of which storing more magnetic energy than the whole HERA ring) with an equal number of circuit breakers and dump resistors. The focusing and the defocusing quadrupoles are powered independently from the dipoles.

8.6   Quench performance of practical magnets

It cannot be taken for granted that superconducting magnets reach the critical current of the superconductor, the so-called 'short-sample' limit[8]. In the past premature quenching and excessive training have been quite common, see the book by M.N. Wilson (1983) for a discussion. Accelerator magnets can fortunately be built so well that the critical field is achieved with little if any training. There are two prerequisites for such an optimal performance:

- the Rutherford-type cable features excellent mechanical stability with good fixation of the strands and the Kapton or Kapton-glass insulation ensures helium porosity;

- the coil and collar structure provides a strong clamping of each winding turn.

The SSC dipoles with 50 mm inner bore and the RHIC dipoles exhibit very little training and feature an ample safety margin. The same applies for the HERA magnets. All dipoles passed the nominal field of 4.7 T at the first attempt and 93% exceeded 5.6 T without quench. For the majority zero or one training step sufficed to arrive at the critical field of 6 T at the test temperature of 4.75 K.

The LHC magnets with a design field of 8.4 T enter a new regime. The magnetic forces and the internal pre-stress are about a factor af two larger than in 5 – 6 T magnets and important structural materials like Kapton or even the soft copper matrix in the cable are close to their plastic limit. However, the quench performance achieved with prototypes leaves no doubt that dipoles of the LHC type can be operated reliably and with sufficient safety margin. Recently a magnet was excited to 9.7 T without quench (R. Perin, private communication). A very remarkable performance has been achieved with a Nb_3Sn model dipole built by Twente University and Dutch industry (den Ouden et al. 1994, 1995). The magnet reached 11 T without training quench. However, a strong ramp rate sensitivity was observed, the quench current dropped almost by a factor of two for ramp rates above 50 A/s. Owing to the small heat capacity of the epoxy-impregnated coil, a low threshold for beam-loss induced quenches has to be expected.

R.B. Palmer (1992) has made an interesting comparison of the training behaviour of a large variety of accelerator dipoles and quadrupoles. He argues that for stability the copper contained within the fine matrix of NbTi filaments should not be counted

---

[8]Critical current measurements are usually performed on short strand or cable samples of 0.1 to 1 m length.

since it is probably degraded by diffusion from the superconductor and moreover its dimensions are smaller than the mean free path of phonons and electrons. Hence the stabilizing effect is provided mainly by the central copper core of the strands and the copper sheath surrounding the filamentary region. An instability parameter is defined

$$\alpha_{inst} = J_{Cu}^2 / R_{SV} \qquad (8.24)$$

where J_{Cu} is the current density in the stabilizing fraction of the copper (at the quench current of the magnet) and R_{SV} the ratio of strand surface exposed to helium to copper volume heated by Ohmic loss. The average number n_q of quenches is plotted in Fig. 8.12 as a function of α_{inst}. A universal trend is observed: n_q increases rapidly with growing α_{inst}. A rough fit is given by

$$n_q \approx \left(\frac{\alpha_{inst}}{0.12\,(\mathrm{kA})^2/\mathrm{mm}^3} \right)^{2.5}.$$

The data along the dashed line are from tests of SSC cable samples (Ghosh et al.

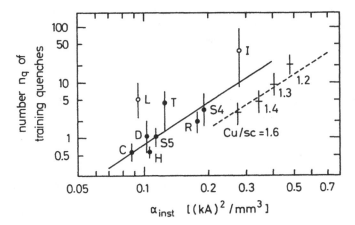

Figure 8.12: The number n_q of training quenches in dipoles as a function of the instability parameter α_{inst}. Experimental data from: CBA (C), DESY prototypes (D), HERA (H), Isabelle (I), LHC prototypes (L), RHIC (R), SSC (S4: 4cm-bore, S5: 5cm-bore), Tevatron (T). The data along the dashed line are from tests on cables samples with different copper-to-superconductor ratios. (After Palmer 1992).

1989) with different copper-to-superconductor ratios. The number of quenches rises from 2 at Cu/SC=1.6 to more than 10 at Cu/SC=1.2 which underlines the detrimental effect of too small a copper proportion in the conductor. It should be noted though that the cable samples came from different manufacturers and were produced at different times. It cannot be excluded that the production process has also an influence on the training behaviour.

In conclusion, we want to emphasize that quench protection is an integral part of the magnet and systems design and cannot be added on afterwards. The first aim is to build magnets with high inherent stability. This implies sufficient copper stabilization of the conductor, good clamping of the windings in the coil and helium transparency. The current connections between adjacent magnets should be reinforced by copper (about 100 mm^2) which can carry the total current for at least the decay time of some 20 s. A good high voltage insulation of the coils is essential. Internal solder joints should have a low resistance (10^{-9} Ω or less) and good cooling.

Quenches in a proton accelerator will happen for many reasons, in particular if a sizeable fraction of the beam ist lost due to malfunctioning power supplies, kicker magnets and other 'standard' components. The quench detection and protection system has to be designed to cope with such events and to function with a high degree of reliability and redundancy. For slow ramping machines, the internal-bypass concept with 'cold' diodes has a definite advantage over the external-bypass concept with thyristor switches since the diodes take over the current from the quenching magnet automatically without requiring a trigger from external electronics and computers. Quench heaters fired by active electronics may provide extra safety in spreading the energy over a large part of the quenched coil.

References

D.E. Baynham, V.W. Edwards and M.N. Wilson, *Transient stability of high current density superconductor wires*, IEEE Trans. **MAG-17** (1981) 732

I.M. Bolotin et al., *The quench detector on magnetic modulator for the UNK quench protection system*, Supercollider 4, J. Nonte Ed., Plenum Press, New York 1992

D. Bonmann et al., *Investigations on heater induced quenches in a superconducting test dipole coil for the HERA proton accelerator*, DESY report HERA 87-13 (1987)

W.H. Cherry and G.I. Gittelman, *Thermal and electrodynamic aspects of the superconducting transition process*, Solid State Electronics 1 (1960), p. 287

J. Chikaba et al., *Relation between instabilities and wire motion in superconducting magnets*, Cryogenics **30** (1990) 649

J. G. Cottingham, *Magnet fault protection*, Brookhaven report BNL-16816 (1971)

A. Devred, *General formulas for the adiabatic propagation velocity of the normal zone*, IEEE Trans. **MAG-25** (1989) 1698

A. Devred, M. Chapman et al. (a), *Quench characteristics of full-length SSC R&D dipole magnets*, Proc. Cryog. Eng. Conf., Los Angeles 1989, Adv. Cryo. Eng. **35**, p. 599

A. Devred, M. Chapman et al. (b), *Investigation of heater induced quenches in a full length SSC R&D dipole*, Proc. 11th Int. Conf. on Mag. Techn. MT-11, Tsukuba 1989, Vol. 1, p. 91

A. Devred, *Quench Origins* in: M. Month and M. Dienes (Eds.), *The Physics of Particle Accelerators*, American Inst. of Physics Conf. Proc. 249, (1992) p. 1262

L. Dresner et al, *Report on the analysis of the large propagation velocities observed in the full-length SSC dipoles*, SSC report SSCL-322 1990

L. Dresner, *On the connection between normal zone voltage and hot spot temperature in uncooled magnets*, Cryogenics **34** (1994) 111

L. Dresner, *Stability of Superconductors*, Plenum Press, New York, London, 1995

A. K. Gosh et al., *Training in test samples of superconducting cables for accelerator magnets*, IEEE Trans. **MAG-25** (1989) 1831

D. Hagedorn and F. Rodriguez-Mateos, *Modelling of the quenching process in complex superconducting magnet systems*, IEEE Trans. **MAG-28** (1992) 366

D. Hagedorn and L. Coull, *Cascade and multiple quenching in the half octant of the LHC*, CERN internal note AT-MA 94-93 (1994)

K. Koepke, *TMAX program*, Fermilab 1980

D. Leroy et al., *Quench observation in LHC superconducting one meter long dipole models by field perturbation measurements*, IEEE Trans. **ASC-3** (1993) 781

Y.M. Lvovsky and M.O. Lutset, *Transient heat transfer model for normal zone propagation*, Cryogenics **22** (1982) 639

G. B. J. Mulder et al, *Quench development in superconducting cable having insulated strands with high resistive matrix (part 2)*, IEEE Trans. **MAG-28** (1992) 739

K.-Y. Ng, *Minimum propagating zone of the SSC superconducting dipole cable*, Internal report SSC-180 (1988)

H. Nomura et al., *Acoustic emission in a composite copper NbTi conductor*, Cryogenics (1980) 283

T. Ogitsu et al., *Quench antenna for superconducting particle accelerator magnets*, IEEE Trans. **MAG-30** (1993) 2773

T. Ogitsu, *Influence of cable eddy currents on the magnetic field of superconducting particle accelerator magnets*, PhD thesis, SSC report SSCL-N-848 (1994)

T. Ogitsu, Ganetis et al., *Quench antennas for RHIC quadrupole magnets*, 14th Int. Conf. on Magn. Techn. (MT-14), Tampere, Finland 1995

U. Otterpohl, *Untersuchungen zum Quenchverhalten supraleitender Magnete*, DESY report HERA 84/05 (1985)

A. den Ouden et al., *An experimental 11.5 T Nb$_3$Sn LHC type of dipole magnet*, IEEE Trans. **MAG-30** (1994) 2320, and: *The Nb$_3$Sn dipole project at the University of Twente*, internal report 1995

R. B. Palmer, *Superconducting accelerator magnets: A review of their design and training*, Int. Conf. High Energy Physics, Dallas 1992, AIP Conf. Proc. 272, p. 1985

D. Perini and F Rodriguez-Mateos, *Calculation of the thermo-structural transient in LHC dipole at quench*, IEEE Trans. **MAG-28** (1992) 370

S. Pissanetzky and D. Latypov, *Full featured implementation of quench simulation in superconducting magnets*, Cryogenics **34** (1994) 795

A. Siemko et al., *Quench localization in the superconducting model magnets for the LHC by means of pick-up coils*, to be published in IEEE Trans. Magn.

Z. J. Stekly and J. L. Zar, *Stable superconducting coils*, IEEE Trans. **NS-12** (1965) 367

J.C. Theilacker, B.L. Norris and W.M. Soyars, *Tevatron quench pressure measurement*, Adv. Cryog. Eng. **39** (1994) 469

T. Tominaka et al., *Quench analysis of superconducting magnet systems*, IEEE Trans. **MAG-28** (1992) 727

M. N. Wilson, *Computer simulation of the quenching of a superconducting magnet*, Rutherford High Energy Laboratory internal report RHEL/M 151 (1968)

M. N. Wilson, *Superconducting Magnets*, Clarendon Press, Oxford 1983

Further reading:

V. A. Al'tov, V. B. Zenkevich, M. G. Kremlev and V. V. Sychev (ed.), *Stabilization of Superconducting Magnetic Systems*, Plenum Press, New York, London 1977

P. Genevey et al., *Cryogenic test of the first two LHC quadrupole prototypes*, IEEE Trans. **ASC-5** (1995) 202

D. Leroy et al., *Test results on 10 T LHC superconducting one metre long dipole models*, IEEE Trans. **ASC-3** (1993) 614

K.K. Leung and Q.S. Shu, *Eddy current and quench loads response for the SSC 4-K liner and bore tube during collider magnet quench*, Adv. Cryog. Eng. **39** (1994) 771

R. G. Mints and A. L. Rakhmanov, *Critical state stability in type II superconductors and superconducting-normal-metal composites*, Rev. Mod. Phys. **53** (1981) 551

W. Nah et al., *Quench characteristics of 5-cm-aperture, 15-m-long SSC dipole magnet prototypes*, IEEE Trans. **ASC-3** (1993) 658

S. L. Wipf, *Stability and degradation of superconducting current-carrying devices*, Los Alamos Scientific Laboratory Report 7275 (1978)

Chapter 9

Impact of Field Errors on Accelerator Performance and Correction Schemes

9.1 Basic elements of accelerator physics

In this section we want to introduce a few important concepts of accelerator physics which are needed to understand the impact of magnetic field errors on the performance of a superconducting storage ring. For a systematic treatment we refer to the literature (Bryant and Johnsen 1993, Edwards and Syphers 1993, Wiedemann 1993, Wille 1992, Rossbach and Schmüser 1994). Circular accelerators are equipped with dipole magnets for bending and quadrupoles for focusing. Quadrupoles have the unpleasant feature of focusing in one plane only but defocusing in the orthogonal plane. An overall focusing in both the horizontal and the vertical planes is achieved by arranging F (horizontally focusing) and D (horizontally defocusing) quadrupoles in alternating order. The hadron and electron accelerators for high energy physics are usually composed of a large number of identical 'FODO' cells, each comprising an F quadrupole, a non-focusing element O (dipole[1] or drift space), a D quadrupole and another element O. The particles conduct quasi-harmonic transverse oscillations about the nominal orbit, called *betatron oscillations*, which are depicted in Fig. 9.1a. The path length along the nominal orbit is denoted as s. For the horizontal displacement $x(s)$ from the design orbit the oscillation can be written as

$$x(s) = a\sqrt{\beta(s)} \cos(\phi(s) + \phi_0) \quad \text{with} \quad \phi(s) = \int \frac{ds}{\beta(s)} . \tag{9.1}$$

In this equation an important accelerator function appears, the so-called *beta function* which has the dimension of a length and is usually quoted in metres. Equation (9.1) shows that the oscillation is amplitude-modulated with $\sqrt{\beta(s)}$ and that the phase advances inversely proportional to the beta function. The constant a (dimension $\text{m}^{1/2}$)

---

[1] The weak horizontal focusing provided by the dipoles can be neglected in a large machine.

and the phase ϕ_0 are given by the initial conditions and can be chosen arbitrarily. The number of betatron oscillations per revolution is called the Q value in European and the tune ν in American nomenclature.

$$Q \equiv \nu = \frac{1}{2\pi} \oint \frac{ds}{\beta(s)} . \qquad (9.2)$$

The betatron frequency is then $f_\beta = Q \cdot f_0$ where f_0 is the revolution frequency of the particles. The envelope of a particle beam is given by the equation

$$E(s) = \sqrt{\varepsilon \, \beta(s)} . \qquad (9.3)$$

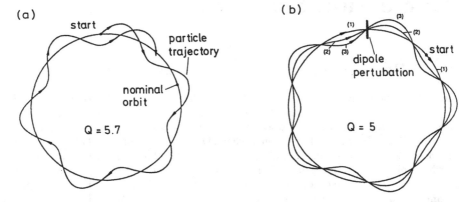

Figure 9.1: (a) Schematic view of horizontal betatron oscillations with non-integer tune (here $Q = 5.7$). The amplitude of the oscillation is grossly exaggerated, it amounts to a few millimetres for a ring of several kilometres circumference. (b) Resonance-like build-up of oscillation amplitude for $Q = 5$. On the first turn the particle is assumed to travel exactly on the nominal orbit.

The quantity ε is called the *emittance*. In a proton machine the emittance ε is a constant of motion when the particle energy is kept constant and it shrinks inversely proportional to the particle momentum during acceleration[2]. The beam has its maximum width (and minimum height) in the centre of the F quadrupoles and its minimum width (maximum height) inside the D quadrupoles. In Fig. 9.2 the horizontal and vertical beam envelopes in a regular FODO lattice are plotted

---

[2]To simplify notation we have considered here only the motion in the horizontal plane, described by the horizontal displacement $x(s)$ from the nominal orbit. The corresponding equations hold for the vertical displacement $y(s)$ from the orbit. So one has to distinguish the horizontal beta function $\beta_x(s)$ and the vertical beta function $\beta_y(s)$ and similarly the horizontal and vertical phase functions $\phi_x(s)$, $\phi_y(s)$, the tunes Q_x, Q_y and the emittances ε_x, ε_y.

as a function of the arc length s along the nominal orbit. Also shown is a particle trajectory in both projections. In the HERA proton storage ring the maximum beam extension amounts to about ± 10 mm at 40 GeV and ± 3 mm at 820 GeV. The pair (Q_x, Q_y) of horizontal and vertical tunes is called the *working point* of the machine.

While the beta function possesses the periodicity of the magnet lattice, this is not at all the case for the particle trajectories. On the contrary, it is strictly forbidden that Q_x or Q_y assume integer values, i.e. that the oscillation closes upon itself after one revolution. If Q is made to approach an integer n the beam is immediately lost. This can be understood as follows. There are always some small dipole errors in the machine, for instance if one of the dipoles is slightly longer than the others. If we choose $Q = n$, a particle traverses such a perturbation with the same betatron phase at each passage. Hence the error kicks received at the perturbation add up coherently from turn to turn, as indicated in Fig. 9.1b. The result is a resonance-like growth of the oscillation amplitude, and after a few turns the particle hits the vacuum chamber. If however Q is sufficiently far away from an integer value the oscillation phase at the position of the dipole pertubation changes from turn to turn and the error kicks average to zero. In a similar way one can demonstrate that half-integer Q values are forbidden because a quadrupole error, however small it might be, would excite a half-integer resonance with rapid beam loss.

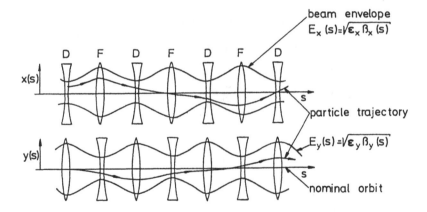

Figure 9.2: Horizontal and vertical view of the beam in a regular FODO lattice. The beam envelopes $E_x(s)$ and $E_y(s)$ are periodic while an arbitrary particle trajectory $(x(s), y(s))$ does not possess the periodicity of the magnet lattice.

There are further restrictions on the betatron frequencies. The first comes from the sextupole magnets which forbid the use of third-integer tunes $Q = n/3$. Sextupoles are needed in the accelerator to correct for the momentum dependence of the focal power of the quadrupoles, the so-called 'chromatic errors' (this name has been chosen in analogy with light optics). Particles with higher than nominal momentum, $p > p_0$, receive less focusing and have therefore a smaller Q value while particles with $p < p_0$ have $Q > Q_0$. The *chromaticity* Q' (often also called ξ) is defined as the variation of tune with relative momentum change

$$Q' = p_0 \cdot \frac{dQ}{dp} \ . \tag{9.4}$$

The momentum deviation changes also the nominal orbit of the particles. This is obvious in a cyclotron where a higher-momentum particle travels on a circle of larger radius. In an accelerator with strong focusing by quadrupoles, an off-momentum particle conducts its betatron oscillation about an orbit which is displaced from the central design orbit by

$$x_D(s) = D(s) \cdot (p - p_0)/p_0 \tag{9.5}$$

where $D(s)$ is the so-called *dispersion function*. The displacement amounts to 1–2 mm typically.

The 'natural chromaticity', produced by the quadrupoles in the arcs of the accelerator, is $Q'_{\text{nat}} \approx -40$ for a ring in the size of HERA. Significant additional contributions come from the very strong quadrupoles at the interaction points. For a typical momentum band in a proton beam of $\Delta p/p_0 = \pm 10^{-3}$ the resulting tune spread would be far bigger than tolerable. A compensation is possible by means of sextupoles which are placed close to the main quadrupoles and act as momentum-dependent quadrupole correctors. However, as mentioned above, they have the unpleasant side effect of exciting third-integer resonances.

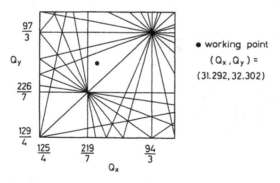

Figure 9.3: Tune diagram of the HERA proton storage ring with resonance lines up to seventh order. The working point is indicated. We thank B. Holzer for providing this figure.

Without magnetic field errors, the integer, half-integer and third-integer resonances would constitute the only forbidden lines in the two-dimensional (Q_x, Q_y)

tune diagram (Fig. 9.3). Unfortunately, resonances of much higher order play a role, driven by the higher-order multipole errors in the superconducting magnets. Moreover, there are magnetic fields present like skew quadrupoles (term a_2) that couple horizontal and vertical betatron oscillations and lead to coupling resonances. Skew quadrupoles arise in dipole magnets from of a top-bottom asymmetry in the coil; another source is a misalignment of the field angle of the main quadrupoles. When coupling is present the resonance condition reads

$$m \cdot Q_x + n \cdot Q_y = p \qquad (9.6)$$

where m, n, p are integer numbers.

The tune diagram of HERA is depicted in Fig. 9.3 together with the resonance lines of order up to seven which must be avoided to prevent particle loss or beam blow-up. In reality the resonance lines are not infinitely narrow but have a certain width which is proportional to the strength of the corresponding multipole field. The working point must stay outside these resonance bands, and from Fig. 9.3 it is obvious that the permitted region for (Q_x, Q_y) shrinks more and more the larger the field errors in the magnets become. This underlines very clearly the importance of high field quality that was addressed in Chap. 5. The field perturbations vanish on the axis of a magnet and grow with r^{n-1} where n is the multipole order. Particles with small betatron amplitudes are therefore not affected but the detrimental effect of high-order multipoles on the stability of particle motion grows rapidly with increasing betatron oscillation amplitude. If we choose the reference radius r_0 of the multipole expansion to be equal to the maximum conceivable deviation of the particle trajectories from the magnet axis (see Sect. 4.3), then a tolerable upper limit for higher-order multipoles is typically $\leq 10^{-4}$ of the main field at r_0.

A very important property of a storage ring is the *dynamic aperture* which can be loosely defined as the maximum 'stable' oscillation amplitude ('stable' means that the particles survive the foreseen storage time without hitting the vacuum chamber). In a superconducting machine the dynamic aperture is generally smaller than the mechanical aperture which is provided by the beam pipe or by collimators.

The most critical limitation arises at injection energy where the beam has its largest extension and where the relative field errors due to persistent currents are maximum. A careful compensation of the persistent-current fields, in particular the sextupole component in the dipoles, is mandatory for achieving a high beam intensity and luminosity in the machine. In HERA also the 12-pole in the quadrupoles requires compensation to avoid a factor of 3 loss in dynamic aperture (Brinkmann and Willeke 1991, Willeke and Zimmermann 1991). The dynamic aperture cannot be computed analytically. Very elaborate computer simulations are needed in which particles with different start coordinates are tracked around the ring for $10^5 - 10^6$ turns. The most accurate determination of a dynamic aperture has been recently made for HERA at 40 GeV (Fischer 1995). It is based on the measured individual field errors of all superconducting magnets, including the different time dependencies of the persistent-current sextupoles (compare Fig. 6.9). The measured dynamic aperture is about 15%

smaller than the computed one which implies that most but not all perturbing effects have been taken into consideration in the model. In view of the complexity of a large machine this must be considered an excellent result. Previous determinations have achieved agreement to within a factor of two.

The tracking models are so advanced that the dynamic aperture of a future machine like the Large Hadron Collider can be predicted with reasonable confidence if 'realistic' magnetic field errors are used as an input. The present calculations made at CERN are based on an extrapolation of the measured average and random multipole errors of the HERA and RHIC magnets. With a slight improvement of the mechanical precision in the coils it appears possible[3] to satisfy the LHC requirements on dynamic aperture which derive from the high design luminosity of $L = 10^{34}$ cm^{-2}s^{-1}.

9.2 Impact of persistent currents on accelerator performance

The following discussion concentrates on the HERA proton ring where the persistent-current effects are a factor of four larger than in the Tevatron owing to the low injection energy of 40 GeV as compared to 150 GeV. The persistent-current sextupole, if left uncompensated, would generate a chromaticity five times larger than the natural chromaticity mentioned above. The sextupole field is compensated locally by sextupole correction coils mounted on the beam pipe inside the dipoles. The small error bars of the sextupole data[4] presented in Fig. 6.5 indicate that the compensation can be done quite accurately, and this was indeed verified in the commissioning phase of the HERA machine. The correction coil currents computed from the measured persistent-current data were almost correct and needed only fine tuning.

The first beam test was a particularly striking example of the capabilities of magnetic measurements. It was made with positrons of only 7 GeV since the nominal 40 GeV protons were not yet available. At the corresponding dipole field of 70 mT (coil current 42.5 A) the persistent-current sextupole component would have been two orders of magnitude larger than tolerable if the standard field cycle had been used. To eliminate the sextupole, all magnets were warmed to 20 K to extinguish any previous superconductor magnetization and cooled back to 4.4 K. Then the current loop 0 → 112 A → 42.5 A was performed which resulted in an almost vanishing sextupole (see Fig. 9.4a). The procedure had been tried out on a single magnet in the magnet test facility and proved representative for the entire ring. A similar procedure was used in the first run with 40 GeV protons, this time with the loop 0 → 314 A → 244.5 A. The

---

[3]J. Gareyte and J.P. Koutchouk, private communication.

[4]The persistent-current sextupole coefficient at 0.227 T is $b_3 = (-34.0 \pm 1.5) \cdot 10^{-4}$ for the dipoles made from the Italian LMI superconductor and $b_3 = (-33.0 \pm 1.2) \cdot 10^{-4}$ for the dipoles made from the Swiss ABB (Asea Brown Bovery) superconductor. The rms spread of $\sigma \approx 1.4 \cdot 10^{-4}$ includes the measurement errors, the magnet-to-magnet variation of superconductor magnetization and the spread due to non-reproducibilities in the initializing field cycle.

measured chromaticity indeed proved an almost perfect sextupole cancellation. For the routine operation of an accelerator these procedures are of course not applicable because they require a warm-up of the whole ring. Instead, the sextupole correction coils must be used to compensate the unwanted field distortions.

A typical injection time in a large proton/antiproton accelerator is 30 minutes. During this period the sextupole exhibits a significant drift which requires an adjustment of the sextupole correctors to keep the chromaticity constant. An interesting question is what happens when the acceleration is started. One could imagine that the time-varying dipole field induces new magnetization currents which overwrite the decaying pattern and thus restore the original sextupole.

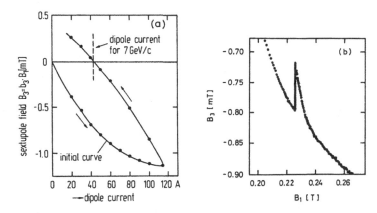

Figure 9.4: (a) Current cycle for the first injection of 7 GeV positrons into HERA. Before performing the cycle, the whole ring was warmed up to 20 K to remove any superconductor magnetization and cooled back to 4.4 K. (b) Drift of the sextupole field B_3 in a HERA reference magnet during the 30-minute injection time at $B_1 = 0.227$ T and re-approach of the hysteresis curve with beginning acceleration (Brück, Degèle et al. 1992).

Fig. 9.4b shows that this is indeed the case. It is seen that the sextupole field B_3 quickly re-approaches the original hysteresis curve. The implication for accelerator operation is that a fast adjustment of the sextupole corrector current is needed. A drift during injection and a re-approach of the hysteresis curve with beginning acceleration is observed also in the dipole field. Here the compensation is done using the horizontally deflecting correction dipoles. The HERA quadrupoles exhibit less drift than the dipoles which might lead to a shift of the working point. Stability is provided by the tune control system.

The time dependence of the sextupole (see Sect. 6.2) translates into a drift of the chromaticity at injection energy. Measured data are shown in Fig. 9.5a. At HERA there exist two 'reference dipoles' (Brück, Degèle et al. 1992), powered in series with the ring magnets, which are equipped with NMR and Hall probes for dipole field measurements and with rotating pick-up coils for determining the time-dependent sextupole field. The measured B_3 variations are used to control the current in the sextupole correctors. Figure 9.5b shows that the horizontal and vertical chromaticities remain constant when this control circuit is operating. Between 40 and 70 GeV (magnet current between 245 and 422 A) the sextupole follows a rather non-linear curve, compare Fig. 6.5. In the commissioning phase, the correction coil current could be varied only linearly between these energies with the consequence of large chromaticity excursions. Also here the measured chromaticities (Fig. 9.6) agreed very well with those predicted from the online B_3 determination in the reference dipoles. These data were taken as the basis of a non-linear modification of the sextupole correction coil current which resulted in a vanishing chromaticity excursion.

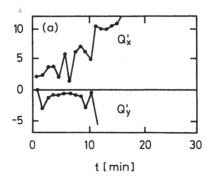 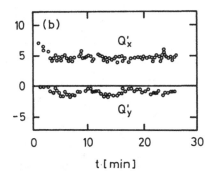

Figure 9.5: (a) Drift of horizontal and vertical chromaticity at the HERA injection energy. The current in the sextupole correction coils is kept constant. (b) Time dependence of chromaticities with the chromaticity control loop in operation. (Holzer 1995).

Another concern is the dependence of persistent current effects on helium temperature. Figure 9.7 shows experimental data on the sextupoles hysteresis curve in a HERA dipole in the range 4.44 to 6.40 K. The temperature of the single-phase helium coolant along the magnet string in HERA varies by at most ± 0.02 K which translates into a negligible variation of the b_3 component of $\pm 0.1 \cdot 10^{-4}$.

The reference magnets play an important role in the ramping procedure of the HERA proton ring. The main superconducting dipoles and quadrupoles are connected in series. The inductance of the superconducting main circuit is 22.14 H while its resistance, given by cables and current leads, amounts to only 37.7 mΩ, leading to a time constant of 590 s. The superconducting correctors and the normal-conducting magnets in the long straight sections of HERA have time constants well below 1 s. To ensure precise tracking of all magnets during acceleration, only the current in the 'slow' main superconducting circuit is raised by computer-generated clock pulses, whereas all remaining magnets and the rf system are controlled by 'dB/dt' pulses which are derived from an induction coil inside one of the reference magnets. The

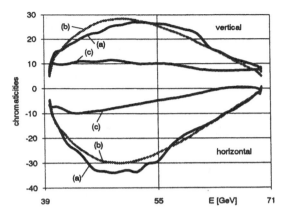

Figure 9.6: Excursion of horizontal and vertical chromaticity during proton acceleration from 40 to 70 GeV in HERA. Curves (a): chromaticities as determined from beam measurements. Curves (b): predicted chromaticities from online measurement of sextupole in reference dipoles. In this experiment the currents in the sextupole correction coils were set by linear interpolation between the 40 and 70 GeV values. In a subsequent run (curves (c)) the sextupole coils were powered according to a non-linear current table. (Meincke, Herb and Schmüser 1992).

number of dB/dt pulses between 0.227 T (40 GeV) and 4.68 T (820 GeV) was found to be reproducible to within 0.001%.

9.3 Correction magnets

9.3.1 Types of correctors

In this section we present a short review of the types of superconducting correction magnets. It has been mentioned in the Introduction that a superconducting hadron accelerator needs the usual correction elements in each cell of the regular FODO lattice:

- dipoles for orbit correction in the horizontal and vertical plane,

- quadrupoles for adjustment of the betatron tunes,

- sextupoles for chromaticity correction and compensation of the persistent-current sextupole components in the main dipoles.

Often octupoles are required at selected locations to provide Landau damping of collective oscillations, and skew quadrupoles serve the purpose of eliminating the coupling of horizontal and vertical betatron oscillations. In the HERA machine,

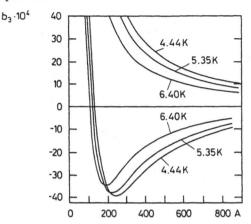

$b_3 \cdot 10^4$

Figure 9.7: Sextupole hysteresis in a HERA dipole for various helium temperatures. The measurements were made by R. Lange of DESY.

owing to the very low injection field of 0.23 T, 10-pole and 12-pole correctors are needed in addition.

The correctors are installed in the long strings of superconducting dipole and quadrupole magnets, hence it is advantageous to use superconducting cables for these magnets as well to avoid an interruption of the cryogenic system.

Superconducting correctors differ considerably in their design from the main magnets. Since many of the correction coils have to be powered individually, a low operating current and consequently a large number of winding turns are desirable. It is then much more difficult if not impossible to achieve the same geometrical accuracy and mechanical pre-stress as in the main magnets. The relative field errors are often an order of magnitude larger but still insignificant in reference to the integrated field of the main dipoles. Fixation of the windings is sometimes accomplished by epoxy impregnation, and for such coils training is quite common. Keeping the operating current well below the critical current of the conductor a satisfactory and reliable performance can be achieved.

Four basic types of correctors have been built (see Fig. 9.8):

- spool-piece coils,

- beam-pipe coils,

- superferric magnets,

- magnets wound from ribbon cable.

The spool-piece concept has been used in the Fermilab Tevatron (Ciazynski, Mantsch 1981). Short dipole, quadrupole and sextupole coils are wound on top of each other,

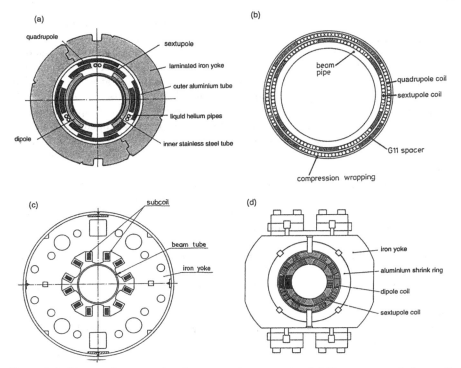

Figure 9.8: Types of superconducting correction magnets: (a) Tevatron spool-piece coils with dipole, quadrupole and sextupole layer; (b) schematic view of the HERA beam-pipe quadrupole and sextupole correctors; (c) RHIC superferric sextupole magnets; (d) LHC dipole/sextupole correctors.

vacuum-impregnated with mineral-filled epoxy and surrounded with an iron yoke made from stamped laminations. Other packages contain quadrupole, sextupole and octupole coils. Some irregularities in the windings were unavoidable. The Tevatron spool-pieces generally exhibited training and could not be excited to the short-sample limit of the conductor; also a mutual influence among the layers was observed. Nevertheless, the spool-pieces have worked reliably for many years of accelerator and proton-antiproton collider operation.

The beam-pipe concept was developed at Brookhaven with the intention of saving maximum space along the accelerator circumference for the main dipole and quadrupole magnets and thereby maximizing the particle energy. Correction windings are mounted on the long beam pipes of the dipoles. A clear disadvantage is the high background field leading to a reduction of current density in the correction windings and large Lorentz forces. A high-performance superconductor is needed. At HERA, the quadrupole and sextupole correction coils (Daum et al. 1989) and the decapole and dodecapole correctors are wound from a single-strand conductor of the type used in the main dipole cable. The wires are insulated with Kapton and glass fibre and then glued onto the insulated beam pipe. The Lorentz forces are taken over by a strong glass-fibre compression wrapping. In an external field of 5 T the coils can be excited to the critical current of the conductor without premature quench. The measured field quality is in perfect agreement with the design, see also Appendix A.

In RHIC a modified version of the pipe correction coil concept is applied for certain types of correctors (Morgillo et al. 1995). The coils are wrapped onto support tubes and concentrically assembled inside an iron yoke. Winding is performed under computer control using ultrasonic power to bond the wire into an epoxy-coated substrate.

Superferric magnets feature a conventional iron yoke and superconducting coils which are more compact than the coils of 'warm' magnets. At HERA the dipole correctors and a number of quadrupoles are built in this manner, at RHIC the sextupoles and trim quadrupoles. If the coils can be orderly wound the quench performance is generally very satisfactory. This applies for the RHIC magnets and the HERA quadrupole correctors. The HERA correction dipoles are of the window-frame type; they are equipped with two saddle-shaped coils with 1000 turns each of a 0.6 mm diameter superconducting wire. Regular winding was not possible, and these magnets indeed exhibited significant training but the specified minimum quench current of 70 A is safely above the nominal operating current of 35 A. The field quality of superferric magnets is good, as can be expected from iron-dominated magnets. Persistent current effects are usually negligible.

A number of correction magnets have been made from a flat cable with parallel unconnected strands that permits winding techniques similar to those used in the main magnets. After winding and mounting of the coils these strands are connected in series. Magnets of this type have been made at Fermilab for the intersection regions of the Tevatron. For the LHC the dipole coil in a prototype dipole-sextupole corrector has been built according to this principle (Baynham et al. 1992, Ijspeert et al. 1993).

9.3.2   Coupled persistent-current effects

The beam pipe correction coils nested inside the main HERA dipole coil give rise to coupled persistent current effects in two ways. Firstly, they act as 'passive' superconductors: any variation in the main dipole field induces magnetization currents in the NbTi filaments which influence the multipole pattern[5]. Secondly, when the correction coils are excited their external field penetrates into the main coil and induces magnetization currents in the dipole coil filaments which have to compete with the already existing persistent currents.

---

[5]It has been suggested by M.A. Green (1987) to use suitably arranged superconducting wires on the beam pipe for a reduction of the persistent-current multipoles of a dipole coil.

Magnetization of correction coil windings by main field

The passive superconductor aspect has been studied experimentally, using the fact that the sextupole and quadrupole correctors extend only over a 6-m section of the 9-m-long dipole coil. In Fig. 9.9 we plot the difference between the sextupole coefficients measured inside and outside this section (at zero correction coil current), and similarly for the decapole. The data agree very well with calculations according to the model described in Chap. 6 and in Appendix C. It is interesting to note that the quadrupole layer, acting as an arrangement of passively magnetized superconductors, generates a sextupole field while the sextupole layer generates a decapole.

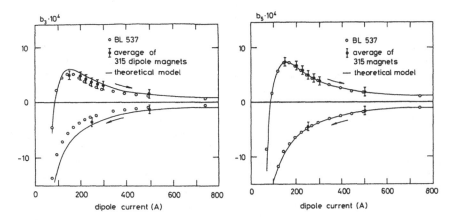

Figure 9.9: Sextupole and decapole components generated by persistent currents in the superconducting wires of the HERA beam pipe correctors, plotted as a function of the main dipole current. The ramp direction is indicated by arrows. Solid points: measured values from 315 dipoles with rms errors; circles: data at from a single dipole; solid curves: model prediction. The correction coil current was zero in these measurements. (Brück, Gall et al. 1990)

Magnetization of the main coil by the correction fields

The 'active' effect of the beam pipe coils is far more complicated, both experimentally and theoretically. Although the fields of the correction coils decrease like $1/r^{n+1}$ ($n = 2$ quadrupole, $n = 3$ sextupole) outside the beam pipe, they are still capable of inducing magnetization currents in the surrounding main dipole coil. These persistent currents counteract the correction coil field at the proton beam. Defining a transfer function f_2 (f_3) as the ratio of field to current in the quadrupole (sextupole) coil, the attenuation has been determined by measuring the transfer function in the normal state at 25 K and in the superconducting state at 4.7 K:

$$f_2(4.7\,\mathrm{K})/f_2(25\,\mathrm{K}) = 0.90\,, \quad f_3(4.7\,\mathrm{K})/f_3(25\,\mathrm{K}) = 0.89\,.$$

This implies that the correctors achieve only about 90% of their nominal strength when their current is ramped up in the superconducting state while keeping the field of the main dipole constant. If however the main dipole field is varied, the attenuating magnetization is overwritten by much stronger magnetization currents due to the dipole field variations. Hence the correctors acquire their full strength. This applies in particular for the acceleration phase of the proton beam where the main dipole field and the correction quadrupole and sextupole fields are raised synchronously. The reduced transfer functions come into play only when the corrector currents are varied to adjust the tune or the chromaticity at a fixed proton energy.

Potentially far more dangerous are higher-order field distortions created during a variation of the corrector current (Pekeler, Schmüser et al. 1992). Figure 9.10 shows a measurement at a dipole field of 0.227 T (corresponding to the HERA injection energy of 40 GeV) in which the current I_2 in the quadrupole coil was moved back and forth between +20 A and −20 A. Surprisingly, the 'unallowed' odd multipole fields B_1, B_3, B_5 are observed and exhibit a rather irregular and highly irreversible behaviour. A theoretical model has been constructed whose basic idea is that the already saturated filaments in the dipole coil are subjected to the time-varying field of the quadrupole correction coil. The model is described in detail in (Pekeler et al. 1992). Here we present only an intuitive picture and list the main results.

Consider two symmetrically arranged NbTi filaments in the main dipole coil, located at $x = \pm R$ and $y = 0$, and assume that they have been completely saturated with persistent currents by a previous field sweep of the dipole field B_1, see state (1) in Fig. 9.11. Now let the quadrupole current be raised from zero to $I_2 = +20$ A. In the right filament at $x = +R$ the field sweep δB_2 has the same direction as B_1 and is therefore unable to induce a new magnetization current as the filament is already saturated. However in the left filament at $x = -R$ the field sweep δB_2 has the opposite direction as B_1 and superimposes a bipolar current of opposite polarity (state (2)). Reducing I_2 to zero again removes these extra currents in the left filament but induces corresponding currents in the right filament (state (3)). So a reversible quadrupole current cycle has resulted in an irreversible change of the magnetization pattern of the two filaments. Going on to $I_2 = -20$ A leads to state (4) and from there to $I_2 = +20$ A finally to state (5). Note that any further variation of I_2 between +20 A and −20 A is equivalent to interchanging states (4) and (5). It is straightforward to compute the even and odd multipole fields generated by the two filaments. The pattern has a striking similarity with the measurements.

For a quantitative analysis all filaments in the dipole coil must be considered. In most of the filaments the current pattern is more complicated than sketched in Fig. 9.11 because the dipole and quadrupole field vectors are in general not simply parallel or antiparallel to each other but include an angle that depends on the position in the dipole coil. The results of the detailed model calculation are plotted in the right-hand part of Fig. 9.10. Very good qualitative and almost quantitative agreement with the data is obtained. This underlines the predictive power of the critical state model.

A similar experiment has been performed in the normal state at $T = 25$ K, see Fig. 9.12, and here the unallowed odd multipole fields are compatible with zero while the allowed field B_2 depends linearly on the current I_2. The complex behaviour displayed in Fig. 9.10 is therefore a truly superconductive phenomenon.

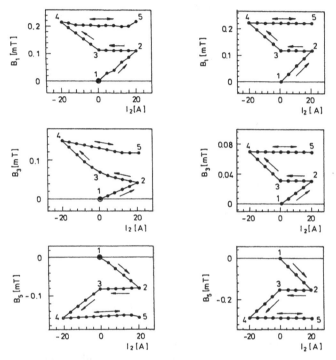

Figure 9.10: Unallowed multipole fields which are generated when the current I_2 in the quadrupole correction coil is varied at a constant dipole field of 0.227 T. Left side: measured dipole, sextupole and decapole fields for the excitation sequence of the correction quadrupole indicated by the points (1) to (5) in the drawing: (1) $I_2 = 0$, (2) $I_2 = +20$ A, (3) $I_2 = 0$, (4) $I_2 = -20$ A, (5) $I_2 = +20$ A. The data were taken at 4.7 K in the superconducting state. Right side: model predictions for the same excitation sequence. When point (5) has been reached, any further variation of the current I_2 in the range ± 20 A leaves the odd multipole fields basically invariant, both in the experiment and in the theoretical model. (Pekeler, Schmüser et al. 1992).

The unexpected 'forbidden' multipoles and their complicated hysteretic behaviour are a consequence of the inherent hysteresis of hard superconductors and the superposition of magnetic fields of different parity (dipole $n = 1$, quadrupole $n = 2$). When

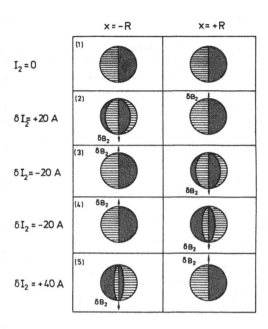

Figure 9.11: Current pattern in two selected dipole-coil filaments for the excitation sequence of Fig. 9.10.

the sextupole correction coil is excited, having the same parity as the dipole, in fact only odd multipoles are observed (Pekeler et al. 1992).

The correction coil currents used in this study exceed the values needed at 40 GeV ($B_1 = 0.227$ T) by almost an order of magnitude, hence in the HERA machine the undesirable effects would be smaller but still well beyond the tolerable limit. Fortunately these field distortions can be completely eliminated by a sufficiently large current sweep of the main dipole. The basic explanation is that the varying dipole field induces new magnetization currents which overwrite the previous current pattern. The following procedure leads to good field quality and has been adopted for HERA: at the end of a luminosity run at 820 GeV (4.68 T) the beam is dumped and then the correction coils are set to their nominal values at 40 GeV. After that the main field is cycled (4.68 T → 0.05 T → 0.227 T), and this large dipole field sweep extinguishes any previous persistent-current multipoles. Some slight variations in quadrupole or sextupole current are still needed for tune or chromaticity adjustments at constant dipole field, however these changes are far too small to cause any harm. So, in summary, the adverse effects of coupled persistent currents can be entirely avoided in practice.

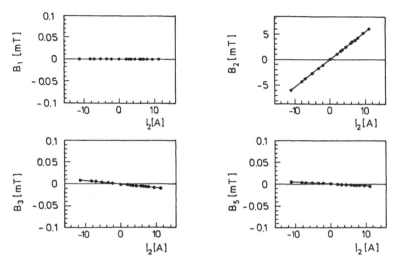

Figure 9.12: Measured dipole, quadrupole, sextupole and decapole fields for the following sequence of correction quadrupole current I_2: $0 \rightarrow +10$ A $\rightarrow 0 \rightarrow -10$ A $\rightarrow 0 \rightarrow +10$ A. These data were taken at 25 K in the normal state for dipole field zero.

Coupled persistent current effects should also be present in other nested superconducting coils and have in fact been recently observed in the LHC dipole/sextupole correctors. For the Tevatron spool pieces no measurement has been made according to our knowledge.

References

E. Baynham et al., *Construction and test of a model of the LHC superconducting corrector magnet MDSBV*, IEEE Trans. **MAG-28** (1992) 354

R. Brinkmann and F. Willeke, *Persistent current field errors and dynamic aperture of the HERA proton ring*, DESY report HERA 88-08 (1988)

H. Brück, D. Degèle et al., *Reference magnets for the superconducting HERA proton ring*, Int. Conf. on High Energy Acc., Hamburg 1992, World Scientific 1993, p. 614

H. Brück, D. Gall et al., *Persistent currents in the superconducting HERA magnets and correction coils*, DESY report HERA 90-11 (1990)

P.J. Bryant and K. Johnsen, *The Principles of Circular Accelerators and Storage Rings*, Cambridge University Press, Cambridge 1993

D. Ciazynski and P. Mantsch, *Correction magnet packages for the Energy Saver*, IEEE Trans. **NS-28** (1981) 3275

C. Daum et al., *The superconducting quadrupole and sextupole correction coils for the HERA proton ring*, DESY report HERA 89-09 (1989)

D.A. Edwards and M.J. Syphers, *An Introduction to the Physics of High Energy Accelerators*, John Wiley, New York 1993

W. Fischer, *An experimental study on the long-term stability of particle motion in hadron storage rings*, PhD Thesis, Universität Hamburg 1995

M.A. Green, *Control of the fields due to superconductor magnetization in the SSC magnets*, IEEE Trans. **MAG-23** (1987) 507

B. Holzer, *Impact of persistent and eddy currents on accelerator performance*, Lecture at the CERN—DESY School on 'Superconductivity at Particle Accelerators', Hamburg 1995, to be published.

A. Ijspeert et al., *Test results of the prototype combined sextupole-dipole corrector magnet for the LHC*, IEEE Trans. **ASC-3** (1993) 773

O. Meincke, S. Herb and P. Schmüser, *Chromaticity measurements in the HERA proton storage ring*, Eur. Part. Acc. Conf., Berlin 1992, Edition Frontières (1992) 1070

A. Morgillo et al., *Superconducting 8 cm corrector magnets for the Relativistic Heavy Ion Collider (RHIC)*, Part. Acc. Conf. Washington D.C. 1995

LHC: The Large Hadron Collider, Conceptual design report, CERN/AC/95-05 (LHC) 1995

M. Pekeler, P. Schmüser et al., *Coupled persistent current effects in the HERA dipoles and beam pipe correction coils*, Proc. Int. Conf. on High Energy Acc., Hamburg 1992, World Scientific 1993, p. 635

M. Pekeler et al., *Coupled persistent current effects in the HERA dipoles and beam pipe correction coils*, DESY report HERA 92-02 (1992)

J. Rossbach and P. Schmüser, *Basic Accelerator Optics*, CERN Accelerator School, CERN report 94-01 (1994)

H. Wiedemann, *Particle Accelerator Physics*, Springer, Berlin 1993

K. Wille, *Physik der Teilchenbeschleuniger und Synchrotronstrahlungsquellen*, Teubner, Stuttgart 1992

F. Willeke and F. Zimmermann, *The impact of persistent current field errors on the stability of the proton beam in the HERA proton ring*, Proc. IEEE Particle Acc. Conf., San Francisco 1991, p. 2483

Chapter 10

Construction Methods of Superconducting Accelerator Magnets

10.1 Overview

10.1.1 Coil

The superconducting coil is the most critical component of a magnet and a sound design is the prerequisite for achieving a high field level without training and a good field quality throughout the whole current cycle. The presently favoured design has evolved over the past two decades. The basic principles stem from the dipoles and quadrupoles of the Fermilab Tevatron (Cole et al. 1979). A number of improvements have been made since at LBL, BNL, DESY, Fermilab, KEK, Saclay, SSCL and CERN. The dipole coils are all based on a suitable approximation of the $\cos\phi$ winding configuration. Since the conductor arrangement has been discussed at length in Chap. 4 we concentrate in the following on important details of the design and some practical aspects of coil production.

10.1.2 Tooling

The demanding task of fabricating 6-m or even 15-m-long magnets with cross sectional accuracies in the order of a few hundredths of a millimetre was first solved at Fermilab with the introduction of *laminated tooling*. The basic idea is that the coil has to conform with such tight tolerances at any cross section whereas in longitudinal direction the requirements are much relaxed. Attempts to produce solid mandrels and molds for coil winding and curing by standard machining techniques had to be given up because they turned out too costly and/or did not comply with the required precision. Precise laminations can be punched at moderate cost and are then assembled to long units of tooling. Laminated tooling has the additional advantage that the mandrels and molds used for short prototypes are identical in cross section to those of series magnets, so the field quality and quench performance results obtained with prototypes are representative for full-size series magnets.

10.1.3   Collars

The coils are surrounded with clamps or 'collars' providing the precise coil geometry and, most importantly, the large pre-stress in the coil needed for good performance at high field. The collars have to meet the same stringent criteria as the tooling and are made from stamped laminations too. The material is usually stainless-steel but great care must be taken that material does not become magnetic by welding, cold work (e.g. during the stamping procedure) or upon cool-down to liquid-helium temperature. Only a few steel types are suitable, for instance 316 LN, Nitronic 40 and DIN 1.4429. In addition to the steel type the chemical composition has to be specified. So-called δ-ferrites may be present in stainless steels which are normally converted to austenite by annealing but may re-appear after cold work or welding. For the HERA dipole collars an aluminium-alloy with high yield strength was chosen (Al Mg4.5 Mn with $\sigma_{02} = 270$ MPa) which is totally non-magnetic (Kaiser 1986). Further details on collar material properties can be found in Appendix G.

In the Tevatron and HERA dipoles the collars are sufficiently stiff to sustain the huge magnetic forces. For the SSC magnets with their larger field this principle would imply rather bulky collar sizes. Here the design has been based on a slim collar that is elastically deformed by the pre-compressed coil but forced into the design shape by the very sturdy iron yoke (Devred et al. 1992). The magnetic forces are largely taken by the yoke. In this type of design the interface between collar and yoke is very critical and particular attention has to be paid to the different thermal shrinking of coil, stainless-steel collar and soft-iron yoke. Owing to the smaller thermal expansion coefficient of soft-iron a certain loss of pre-stress upon cooldown has to be accepted.

The approach at Brookhaven for the RHIC magnets has been different (Thompson et al. 1991). Non-magnetic clamps are avoided but rather the iron yoke itself is used to compress the coil. A precise glass-phenolic form piece serves as a spacer between coil and yoke and also as an excellent electrical insulator. The spacer is produced by injection-molding and the resin is mineral-loaded to reduce shrinkage upon cooldown. The disadvantage of unfavourable differential shrinkage (see Sect. 5.5) and loss of pre-stress during cooldown remains but creates no problems for the moderate field levels in RHIC.

10.1.4   Iron yoke

The main purpose of the iron yoke is to screen the fringe field outside the magnet to an acceptable value of 10 mT or less. The field enhancement provided by the yoke and its influence on field quality in case of saturation have already been discussed in

Chap. 4. Some further points have to be taken into consideration before a decision on the type of yoke can be made:

- quench protection system of a long string of magnets,

- heat load on the cryogenic system,

- cooldown and warmup times of the accelerator,

- pre-stress in the coils.

With these points in mind, we summarize the relative virtues and drawbacks of the 'warm-iron' and the original 'cold-iron' yoke designs.

'Warm-iron' yoke

Magnets with this type of yoke have two advantages: the 'cold' mass is quite small and the field distortions from iron saturation are almost negligible.

The disadvantages are: the iron contribution to the central field is only in the order of 10%; the coil must be well centred in the yoke to avoid eccentricity forces and field distortions (normal or skew quadrupoles in dipole magnets), so many support planes are needed (typically one per metre length) leading to a large heat load on the liquid-helium system; a passive quench protection system by parallel diodes is not easily possible, it would require a costly parallel helium transfer line.

'Cold-iron' yoke

The advantages of a magnet with the iron yoke inside the cryostat and very close to the coil are: the yoke contributes 35 – 40% to the central field, so a substantial savings in superconductor is possible; the coil is automatically well centred, no eccentricity forces arise; the yoke is a stiff body and only a few support planes are needed (two for a 6-m-long magnet) which reduces the heat load considerably; a passive quench protection system with 'cold' diodes that bypass the magnet coils is easily implemented[1].

The disadvantages are: the 'cold' mass is large; the close proximity between coil and yoke causes strong iron saturation with a non-linear current-field relationship and field distortions; soft-iron shrinks much less upon cooldown than the coil, so a very high room-temperature pre-stress is needed which might be a danger for the superconductor insulation.

---

[1]In a magnet whose iron yoke is inside the liquid-helium container, the input and output current leads of the coil can be located at the same end of the cryostat. Here is also the position of the diode. The superconducting cable which feeds the main current to the next magnet in the accelerator ring is guided through a groove on the outer rim of the iron yoke. The yoke screens the particle beam from the fringe field of the current bus. In a warm-iron magnet this is obviously impossible unless an external cryogenic pipe for the main current bus is provided.

'HERA-type' yoke

In the prototype program for HERA both warm-iron magnets (dipoles at DESY, quadrupoles at Saclay) and cold-iron magnets (dipoles at Brown Bovery, Mannheim) had been built. In 1984 the idea was conceived of a magnet which combined most of the positive features of both development lines while avoiding the more serious drawbacks. Stimulated by a comparison of the quench safety of warm-yoke and cold-yoke magnets (Mess, Schmüser 1984) it was proposed (Balewski et al. 1984) to adopt the aluminium-collared coil of the warm-iron development line and surround it with a cold-iron yoke. Calculations and prototype measurements verified (Wolff 1985) that the 'HERA-type' dipole has indeed very favourable properties: compared to the warm-yoke dipole there is a 12% gain in central field permitting dipole fields of more than 6 Tesla with little saturation; passive protection by parallel diodes is possible; no asymmetry forces exist and the heat load on the 4 K system is modest; the pre-stress in the coil at room temperature is the same as in the warm-yoke design.

The only remaining drawback is the large mass to be cooled to liquid-helium temperature. Experience at Fermilab and DESY has shown that the warmup of an accelerator section for maintenance or repair work is not frequently encountered (Wolff 1993). The smallest cryogenic section in HERA is an octant comprising 52 dipoles and 26 quadrupoles. Cooling from room temperature to 4 K requires about 120 hours; the warm-up to 300 K takes the same time (Clausen et al. 1991). So if the machine is built with the degree of reliability which is needed anyhow for long term storage ring operation, the large 'cold' mass does not seriously affect the accelerator efficiency. (Incidentally, the cooldown or warm-up of a long string of warm-iron magnets would be only slightly faster because the limitation is not so much given by the available cryogenic power but by the requirement that excessive thermal stress in the magnets must be avoided).

Twin-aperture yoke

For proton-proton colliders two magnet rings are needed. The SSC design was based on separate magnets and cryostats while for the LHC the idea of the twin-aperture magnet (Dahl et al. 1985) was adopted, for financial reasons and to save space. Two collared coils of opposite field orientation are put into a common iron yoke and cryostat (Leroy et al. 1988). Most of the magnetic flux returns through the neighbouring coil so the yoke can be made slim at the sides in spite of the high design field of 8.4 T. A clear disadvantage of the 'two-in-one' principle is the loss of left-right symmetry for either coil. The dipoles suffer from normal quadrupole components which may be as large as 2% at high excitation. Special measures are taken to reduce these effects, for instance by using ferromagnetic inserts in the non-magnetic collars. The latest LHC dipole design, shown in Fig. 4.13, features a common aluminium-alloy collar for both coils and a yoke with a separation in the vertical plane (LHC 1995). At room temperature the gap between the half-yokes is open. A stainless-steel shrinking cylinder, surrounding the yoke, closes the gap upon cooldown with sufficient tension such that the gap remains closed up to maximum field. The LHC quadrupoles (Fig. 10.1) are also of the twin-aperture type. The magnetic coupling is so small that the resulting dipole component can be neglected.

10.1.5 Cryostat

The cryostat of an accelerator magnet consists of a helium container housing the coil, a surrounding heat shield at an intermediate temperature, a vacuum vessel with internal tubing and a support structure. The thermal insulation must be very effective because a heat influx of 1 W at 4 K requires almost 300 W of electrical power in the most advanced helium refrigerator. For 2 K operation this number more than doubles. In the Tevatron magnets the cryostat is very slim as it has to fit in between the collared coil and the iron yoke. In cold-yoke magnets ample space is available which simplifies a low heat load design.

The cooling scheme plays an essential role for the detailed layout. Accelerator magnets are generally operated in long strings containing many tens of magnets. The temperature is 4.3 –4.5 K (Tevatron, HERA, RHIC) and 1.9 K in the LHC. Single-phase helium with a pressure above the equilibrium vapour pressure is passed through the string, then expanded in a Joule-Thomson valve where it converts to a two-phase mixture of liquid and vapour, and is finally returned through the string. The two-phase helium, which has almost constant temperature, removes heat influx from the single-phase liquid via heat exchange. The support structure of the coil may consist of simple epoxy-fibreglass blocks (Tevatron), long epoxy-fibreglass bands (e.g. HERA), titanium rods (e.g. in UNK, see Ageev et al. 1992) or epoxy-fibreglass cylinders with extremely low heat conduction (SSC, see Nicol 1992, LHC, RHIC). The heat shield at 50 – 70 K is wrapped with some 30 layers of superinsulation, a thin aluminized polyester foil, which reduces the heat radiation from 300 K to 70 K to about 0.7 W/m². In the LHC magnets a second heat shield at 5 K is required to suppress heat influx into the 2 K superfluid helium system to a few tenths of a watt per magnet. As an example of an advanced cryostat design with very low heat load we show in Fig. 10.1 the cross section of the LHC quadrupole cryostat (Cameron et al. 1994, LHC 1995). For a detailed discussion of cryogenic systems for accelerators we refer to Lebrun (1994).

10.2   Coil fabrication

In the following we describe the fabrication of the HERA dipole coils (Wolff 1985, 1987). Many of the methods have been adopted from the pioneering Tevatron magnets and similar procedures have been or are being used for the SSC, LHC and RHIC magnets.

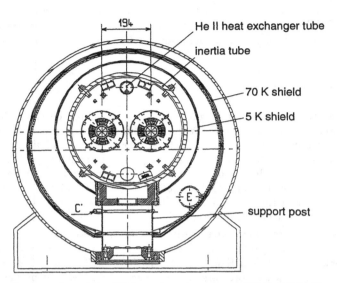

Figure 10.1: Cross section of the LHC quadrupole cryostat (LHC 1995). The quadrupoles were developed at Saclay. The two coils are mounted in a common iron yoke which is installed in a thick-walled stainless steel tube. This 'inertia tube' provides excellent alignment and straightness of the quadrupoles. A similar tube is used in the HERA quadrupoles.

10.2.1   Winding and curing of half-coils

Coil winding

The mandrel for coil winding and the mould for baking are stacked from punched steel laminations providing a geometrical accuracy of 0.02 mm at any coil cross section. The Rutherford cable is wound with an electronically controlled tension of 200 N. First, the inner half-coil comprising 32 turns is wound. It is covered with a mould, transferred into a hydraulic press and heated to 90°C. When the B-stage epoxy in the cable insulation has become soft the coil is compressed to the required shape and afterwards cured at 160°C. The outer half-coil with 20 turns is wound onto the cured inner half-coil and then the whole assembly is baked out again. A 100-mm-long solder joint serves as electrical connection. The resistance of about 10^{-9} Ω is sufficiently low that the produced heat is easily conducted away by the liquid helium. The inner and outer half-coil are separated by a 0.5 mm thick epoxy-fibreglass layer with slots for helium passage.

Coil fabrication must take place in a clean environment, strictly avoiding oil, dust and any kind of chips, especially metallic ones. Foreign particles enclosed in between the turns may damage the insulation during coil compression and produce turn-to-turn shorts. Starting from a spool of insulated cable with a continuous length sufficient for at least one coil layer, the cable is wound around a coil mandrel which is mounted on a steel beam. There exist different concepts for a winding machine. The mandrel may be placed on a long table that is mounted to the floor and the cable spool is moved around on a motor-driven wagon. Alternatively, the cable spool is fixed in a frame connected to the floor and the table with the winding mandrel is moving back and forth underneath the spool in axial direction. For both types of machines the winding mandrel must be rotatable about its longitudinal axis to permit adjustment of the azimuthal angle for each winding layer. The winding tension is accomplished by attaching an electric motor to the spool axis via a magnetic clutch that provides an adjustable torque.

The winding mandrel is made from precisely punched steel laminations. The lamination of the HERA dipole mandrel (Fig. 10.2), has the cross section of a half circle on top of a square. Stacks of 200 mm length are pre-assembled on a precise fixture, using accurate stainless-steel rods for relative alignment of the laminations. The stacks are then connected by longitudinal welds and mounted on a machined steel beam which provides sufficient stiffness. After assembly a steel tube for heating and cooling purposes is pulled through the 30-mm hole of the mandrel and brought into close thermal contact with each lamination by means of a roller burnisher. The groove on top accepts and precisely centers the winding keys for the inner resp. outer coil layer. These keys are again made from stamped laminations and are preassembled to sections of 200 mm length.

Before starting the winding, the mandrel is covered with a 0.2 mm thick stainless steel sheet that is sprayed with mold release to prevent sticking of the epoxy during bakeout of the coil. Steel shims of suitable thickness are attached to the winding keys, and the innermost epoxy-fibreglass pieces of the coil heads are screwed to the mandrel. The winding of an inner half coil starts with the turn adjacent to the key, after the cable ramp to outer coil layer has been prepared. Special care must be taken at the coil heads. The winding tension should be tangential to the surface of the winding mandrel to avoid shear forces that might damage the insulation at the narrow cable edge. The coil heads are equipped with epoxy-fibreglass spacers which are bolted to the mandrel after the appropriate turn and provide longitudinal stability during winding. When the last turn next to the midplane has been finished the cable end used for magnet-interconnection is reinforced with copper. Precise steel bars are bolted to the mandrel with slotted holes allowing for vertical motion. In the curing press these bars compress the coil to its required shape. The half-coil is covered with a 1-mm-thick stainless-steel hood with mold release on its inner surface.

Coil curing

A curing mold, covering the entire length, is placed on top of the coil package (Fig. 10.2) and the assembly is moved into a 10-m-long hydraulic press whose maximum force is about 10^6 N/m. Heat carrier oil is pumped through the heat exchanger tubes

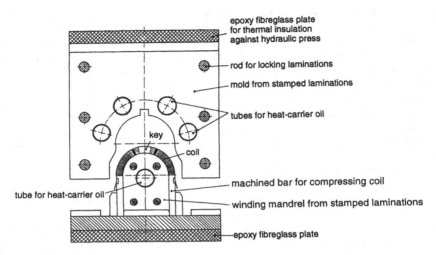

Figure 10.2: The inner half coil of a HERA dipole after compression and curing. For clarity the curing mold has been shifted upwards.

in the mold and the mandrel to heat the assembly for curing of the epoxy. The tube arrangement chosen guarantees a uniform temperature distribution over the whole coil volume which is essential for achieving a uniform softening of the B-stage epoxy and proper alignment of the windings. A further improvement would consist in applying mechanical vibrations during the first heating and pressing period. The final curing temperature is held at 140–160°C for several hours depending on the characteristics of the B-stage epoxy. Then the mold is cooled down by pumping cold oil through the heat-exchanger tubes.

The outer coil layer is wound onto the cured inner coil layer with the slotted epoxy-fibreglass sheet in between. Winding and curing is similar to that of the inner layer. Alternatively the outer half-coil can be wound and cured as a separate unit and then glued on top of the inner half-coil.

Checks for turn-to-turn shorts are performed by connecting the cured half-coil to a charged capacitor to form an LC circuit and measuring the decay time constant of the oscillations. A voltage of more than 40 Volts can be achieved between adjacent turns. The elastic modulus of the coil and the arc length are measured at many locations along each coil leg. This can be done in an automated way by moving a pressing tool along the coil and applying at least two different pressures, the maximum pressure being close to the final pre-compression. Owing to the inhomogeneous structure of cable and insulation one observes a considerable variation of elastic modulus and arc length along the coil. After the local compression the turn-to-turn voltage test is repeated.

10.2.2   Coil ends and solder connections

The coil heads must be carefully designed since the windings cannot be confined as well as in the straight section. Quite often magnets have been found to quench in the end region. In a simple coil head design the cable has to be bent over its narrow edge which is difficult to achieve if the cable width exceeds 10 mm, in particular for inner coil diameters of 50–60 mm (SSC, LHC). The mechanical stability of the coil heads can be improved by inserting precisely shaped spacers between appropriate turns. We have seen in Chap. 4 that spreading the windings in the return region is also desirable in order to remove local field enhancements and to minimize the end-field contribution to higher harmonics. The spacers can be designed in such a way that the two sides of the cable follow curves of equal arc lengths ('constant perimeter' windings), see Fig. 10.3.

Figure 10.3: Coil head region of an LHC dipole and spacers for a 'constant-perimeter' winding configuration as computed with the program ROXIE (courtesy S. Russenschuck).

The inner and outer half-coils of a long dipole are wound from a continuous section of superconducting cable. Internal splices between the half-coil layers are made by soldering. Splices should be located in a region of moderate field, for instance in the outer coil layer or even outside the coil. Good cooling is mandatory and an overlap of the two cables of about ten times the cable width. A suitable solder is Sn 5%Ag or Sn 40%Pb. Before soldering the wires are thoroughly cleaned and oxide layers are etched away. It is important to use non-corrosive flux. The soldering temperature should be controlled to avoid degradation of the superconductor. With these measures a resistance below $\leq 2 \cdot 10^{-9} \, \Omega$ has been achieved at HERA resulting in a total power dissipation of less than 150 mW at 5000 A in the three internal splices of a dipole.

10.2.3   Quench heaters and ground insulation

Most accelerator dipoles are equipped with electric heaters whose purpose is to drive the entire coil into the normal state once a quench has happened. Thereby the stored magnetic energy is dissipated in the entire coil volume and local overheating is avoided. The heaters are fired by discharging capacitor banks. In the Tevatron and HERA dipoles the quench heaters consists of 30 μm thick and 12 mm wide stainless-steel strips glued onto a 125 μm thick Kapton foil which is placed at the outer circumference of the second coil layer. As mentioned earlier, this is by no means the optimal position since the magnetic field is fairly low in this region resulting in a slow quench propagation velocity and a rather high trigger threshold. A far better location would be the first turn of the inner layer where the field is highest, however this solution was not adopted owing to complications in the assembly and the risk of heater damage or turn-to-turn shorts. A detailed study of quench heater design and performance has been presented by Christianson, Fagan and Roach (1994).

The coil has to be carefully insulated against the clamping structure with a material suitable for low temperature and high radiation dose ($\sim 10^6 - 10^7$ Gray). Kapton is one of the few candidates. The insulation to ground must stand at least twice the 'nominal' voltage plus an additional 1 kV. The nominal voltage is defined as the coil-to-ground voltage during a worst-case quench in the magnet system. It can be in the order of 1–1.5 kV.

The breakdown voltage in liquid helium is rather high. However, helium evaporation may occur with a drastic reduction in breakdown voltage, see Appendix E. At room temperature and a pressure of one bar the breakdown voltage in helium is about a factor 5 – 10 lower than in air. Therefore the ground insulation must be designed for a minimum path length of 10 mm between conductor and clamping structure. Kapton foil is particularly vulnerable to pin holes and moreover the ground insulation is heavily squeezed by the collars which bears the risk of damage at sharp corners. For these reasons the ground insulation of the HERA dipoles is made of 6 layers of 125 μm thick Kapton foil with sufficient mutual overlap. The ground insulation of the RHIC magnets is provided by the glass-phenolic form pieces depicted in Fig. 4.4.

10.2.4   Nb$_3$Sn coils

Niobium-tin coils are wound from Rutherford-type cables with the same techniques as applied for niobium-titanium. The essential difference is the sensitivity of the conductor to mechanical deformation. In the 'react and wind' method, the fully reacted cable is used to wind the coil. The superconductor may be heavily degraded by bending the cable around the small radii in the coil-head region. One possibility is to increase the radius of curvature by flaring out the coil ends[2] but the presently favoured technique is to wind the coil from the unreacted cable and fire it afterwards at about 675°C for 140 hours to react the Nb$_3$Sn. The cable insulation consists of heat-resistant glass or glass-mica tape with some organic resin which is burnt during firing. The reacted half-coil is very fragile and must be vacuum-impregnated with epoxy to improve the mechanical stability and prevent superconductor degradation

---

[2]A magnet of this type was built in Brookhaven.

caused by transversal stress. The assembly of finished half-coils and the collaring procedure is similar to the NbTi case. Because of the high operating fields the iron yoke must accept the major part of the magnetic force.

10.3 Collaring of coils

10.3.1 Assembly of dipole coil

The two half-coils of a dipole must be carefully selected to minimize field distortions, especially skew quadrupoles. For the HERA dipoles the strategy has been to optimize the arc length of the half-coils for minimum sextupole and to suppress the quadrupole components (normal and skew) by combining half-coils with properly matched average arc length and elastic modulus. The two half-coils are insulated at the midplane and then placed around an inner mandrel that is needed in the collaring process. The mandrel, which may be rigid or collapsible, has to withstand the collaring forces and must be removable afterwards. In case of a rigid mandrel the contact area to the half-coils should be restricted to the midplane and the limiting angles to reduce friction.

The laminated HERA dipole collar is depicted in Fig. 10.4. A single punched lamination has the shape of a 'half-U'. Collar half-packs of 150 mm length are assembled by flipping every second lamination so that a full 'U' is formed. The half-collar packs are pre-assembled by connecting the aluminium-alloy laminations with stainless-steel rods. The half-collar packs are assembled in alternating order from top and bottom around the insulated coil. The toothlike structure of the half-collars interleaves on both sides of the coil. With aluminium-alloy as a collar material some abrasion was observed, so it turned out necessary to cover the laminations with MoS_2 to reduce friction.

The insulated coil, assembled on the central mandrel and surrounded with the loose collar packs, is put into a U-shaped, precision-machined stainless steel beam providing accurate relative alignment of the individual collar laminations. A second U-shaped beam is placed on top and the whole assembly is moved into a hydraulic press. During compression the distance between the steel beams of the press is monitored. After reaching the final closing position, the hydraulic pressure is released in order to be able to remove the central mandrel, and then it is raised again. The force needed for closing the collars amounts to 3500 tons for the 9-m-long HERA dipole and is about twice as high as the desired pre-compression force, owing to friction between adjacent collar laminations. The half-collar packs are connected by stainless-steel rods that are pushed through the punched holes in the horizontal plane. These holes are slightly oversized and an overpressing by about 0.1 mm is necessary to facilitate insertion of the rods. For the coil ends special collar blocks with a cylindrical inner surface are needed.

Alternatively the half-packs can be connected by longitudinal welds (Tevatron

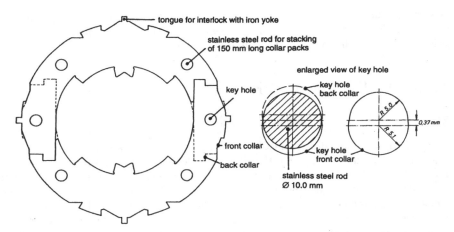

Figure 10.4: Two layers of the HERA dipole collar assembly. The front layer is drawn with continuous lines, the back layer with dashed lines. Shown in an enlarged view is the locking of top and bottom 'U' by longitudinal stainless steel rods of about 9 m length and 10.0 mm diameter. To facilitate insertion of the rods, the holes on the left- and right-hand side of the coil aperture are not circular in cross section but have the shape of a 'key-hole'. The coil is slightly overpressed while the two rods are pushed in. When the force of the hydraulic press is released the coil springs back by 0.1 mm and the two stainless-steel rods interlock the half-collars. A re-opening of a collared coil for improving the field quality by changing the shims is rather easily accomplished.

dipoles) or by tapered keys which are inserted into grooves on the outer circumference (Tevatron and HERA quadrupoles, SSC dipoles).

10.3.2   Coil compression and collar deformation

In the HERA magnet development program the elastic deformation of the clamping structure during collaring, cooldown and field excitation was investigated analytically by G. Meyer (1982) for two collar materials, stainless-steel and aluminium-alloy. The design value of the coil pre-compression was set to 64 MPa at 4 K and zero field; a remaining pre-compression of at least 20 MPa at 5 Tesla was requested. For stainless-steel collars this implies a pre-compression at room temperature of 72 MPa while with aluminium-alloy collars only 46 MPa are needed owing to the stronger thermal contraction of Al. The lower room-temperature pre-stress was one of the main motivations to choose aluminium-alloy as collar material for the HERA dipoles. A disadvantage of aluminium is the small elastic modulus, leading to larger deflections under load than obtained with a stainless-steel collar of equal size. As mentioned in Chap. 5 the effect on higher multipoles is still tolerable. The collar is elastically

deformed by the counter force of the pre-compressed coil, see Fig. 5.10. This can be
utilized to monitor the achieved pre-compression.

10.4   Fabrication and assembly of iron yoke

The yoke is made from low-carbon sheet iron with a low coercive force. The measured
permeability of the iron used in the HERA magnets is shown in Fig. 10.5a. A
histogram of the coercive force is depicted in Fig. 10.5b. The iron may saturate
at high fields. In order to keep the field outside the yoke below a level of 10 mT
the field in the iron should not exceed 1.75 T near the outer surface, a value which
determines the radial thickness of the yoke. The field quality is sensitive to variations
in permeability. Therefore iron with a low spread in magnetic properties is required.

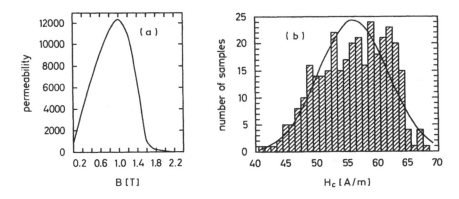

Figure 10.5: (a) Permeability of the HERA yoke iron as a function of field. (b) Distribution
of coercive force. (Sinram 1988)

The dipole yokes usually consist of two halves. The parting should preferably
be in the vertical plane because then no gap has to be crossed by the magnetic
flux. If horizontal separation is chosen (for instance in the SSC prototype magnets
made at Brookhaven, see Goodzeit 1992) care must be taken that the gap closes
completely upon cooldown because small variations in gap width generate sizeable
field distortions. The 5-mm-thick HERA yoke laminations were made by punching
and fine blanking. Stamping may change the permeability in the cutting region but
this is of minor importance if the iron boundary is far from the coil center. The
symmetric half-yoke laminations are stacked on a long steel table with magnetic
attraction to the table surface. Longitudinal compression is necessary to achieve a
uniform filling factor. Near the coil ends the soft-iron laminations are replaced by
non-magnetic stainless-steel laminations. The assembled half-yokes are held together

by long steel rods. One half-yoke is put on a table, the collared coil is inserted and then the second half-yoke is placed on top. The half-yokes are aligned with respect to each other by precise rods inserted into triangular grooves at the parting plane. The yoke/coil assembly is moved into a press and the half-yokes are connected by two longitudinal TIG (tungsten-inert-gas) machine-welds that are made simultaneously. The yoke is surrounded with a stainless-steel tube made from two half shells. The welding of the tube is made in a fixture which defines the radius of curvature of the dipole that is needed in the accelerator ring. The tube serves in addition as the liquid helium container.

In Chap. 5.1 we have seen that the dipole coil has to be precisely centred in the yoke to avoid quadrupole distortions. Groove-and-tongue joints are well suited to interlock collar and yoke but it is advisable to insert a sliding element, for instance a U-shaped bronze sheet, in between the collar and yoke laminations. Otherwise the longitudinal forces during excitation might induce a slip-stick motion of the collared coil inside the yoke. In some early SSC prototypes this was the reason for premature quenches. It is advantageous to transfer the longitudinal forces to thick end plates which are welded to the stainless-steel tube surrounding the yoke.

References

A. I. Ageev et al., *The development and study of superconducting magnets for UNK*, IEEE Trans. **MAG-28** (1992) 682

K. Balewski, H. Kaiser and P. Schmüser, *Ein Kalt-Eisen-Magnet mit Aluminium-Klammer*, DESY internal note, Feb. 1984

W. Cameron et al., *Design and construction of a prototype superfluid helium cryostat for the short straight sections of the CERN Large Hadron Collider (LHC)*, Adv. Cryog. Eng. **39** (1994) 663

O.R. Christianson, T.J. Fagan and J.F. Roach, *High energy booster dipole magnet (HDM) quench heater design and projected performance*, Adv. Cryog. Eng. **39** (1994) 461

M. Clausen et al., *Cryogenic test and operation of the superconducting magnet system in the HERA proton storage ring*, Adv. Cryog. Eng. **37A** (1991) 653

F. T. Cole et al. (eds.), *A report on the design of the FNAL superconducting accelerator*, Fermilab report 1979

P. Dahl et al., *Performance of four 4.5 m two-in-one superconducting R&D dipoles for the SSC*, IEEE Trans. **NS-32** (1985) 3675

A. Devred et al., *Review of SSC dipole magnet mechanics and quench performance*, Super-collider 4, p. 113, Plenum Press New York, London 1992

C. Goodzeit, *Cold mass mechanical design, quench and mechanical test results for full length 50 mm aperture SSC model dipoles built at BNL*, Proc. Int. Conf. on High Energy Accel., Hamburg 1992, World Scientific (1993) 584

H. Kaiser, *Design of superconducting dipole for HERA*, Proc. Int. Conf. on High Energy Part. Accel., Novosibirsk 1986, and DESY report HERA 1986-12

D.C. Larbalestier, *Selection of stainless steel for the fermilab energy doubler/saver magnets*, Fermilab report TM-745 1630.000, 1977

Ph. Lebrun, *Cryogenic systems for accelerators*, Lecture at the US-CERN-Japan Topical Course 'Frontiers of Accelerator Technology', Hawaii 1994

LHC: *The Large Hadron Collider*, Conceptual design report, CERN/AC/95-05 (1995)

D. Leroy et al., *Design of a high-field twin-aperture superconducting dipole model*, IEEE Trans. **MAG-24** (1988) 1373

K.-H. Mess, P. Schmüser, *Vergleich der Quench-Sicherheit bei Warm- und Kalteisen-Magneten*, DESY internal note, Jan. 1984

G. Meyer, *Spannungen und Verformungen der Klammer*, DESY internal note, 1982

T.H. Nicol, *Cryostat design for the Superconducting Super Collider dipole* in: M. Month and M. Dienes (Eds.), *The Physics of Particle Accelerators*, AIP Conf. Proc. **249** (1992) 1230

K. Sinram, *The Influence of fine blanking on the magnetic properties of soft magnetic steel*, IEEE Trans. **MAG-24** (1988) 839

P. Thompson et al., *Revised cross section for RHIC dipole magnets*, Proc. IEEE Part. Acc. Conf., San Francisco 1991, p. 2245

S. Wolff, *The superconducting collared coil for dipoles of the proton ring of HERA - description and fabrication procedure*, DESY internal note 1985, revised 1987

S. Wolff, *Superconducting HERA magnets*, IEEE Trans. **MAG-24** (1988) 719

S. Wolff, *Operational experience with large superconducting magnet systems*, Proc. World Congress on Superconductivity, Munich 1992, Pergamon Press, p. 1457, 1993

Appendix A

Techniques of multipole measurements

A.1 Pick-up coils and determination of multipoles

The most commonly used device for determining the harmonic content of a superconducting dipole or quadrupole is a rotating pick-up coil. Two basic types exist which are sketched in Fig. A.1: a radial coil measures the azimuthal field component while a tangential coil is sensitive to the radial field component. Also 'harmonic' coils are sometimes used to measure a specific multipole field. These coils are often named after Morgan. An example is the sextupole pick-up coil in the reference dipoles of the HERA collider which consists of three tangential subcoils separated by 120°. The dipole field induction cancels in the sum signal.

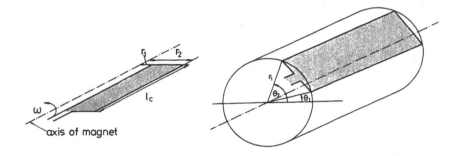

Figure A.1: (a) Radial pick-up coil. (b) Azimuthal pick-up coil.

Here we confine ourselves to the first type and compute the magnetic flux through a radial coil of inner radius r_1, outer radius r_2, length l_c and with N_c turns. If θ is the angle of the coil plane with respect to the horizontal axis the flux is according to

Eq. (4.13)

$$\Phi(\theta) \; = \; N_c \, l_c \int_{r_1}^{r_2} B_\theta(r, \theta) dr$$

$$= \; F \sum_{n=1}^{\infty} \frac{1}{n} \left[\left(\frac{r_2}{r_0} \right)^n - \left(\frac{r_1}{r_0} \right)^n \right] [b_n \cos(n\theta) + a_n \sin(n\theta)] \qquad (A.1)$$

with $F = N_c \, l_c r_0 B_{ref}$. If the coil rotates at a uniform angular velocity ω one gets $\theta = \omega t$ and the induced voltage $U = -d\Phi/dt$ is proportional to the angular velocity, which is difficult to control to the desired precision of 10^{-5}. For this reason the voltage is integrated over the time intervals

$$\Delta t = \frac{T}{N}$$

where $T = 2\pi/\omega$ and N is the number of subdivisions for a complete rotation. This is equivalent to measuring the magnetic flux $\Phi(\theta_k)$ at the discrete angles

$$\theta_k = k \frac{2\pi}{N}, \quad k = 0, \ldots N - 1 \, .$$

The multipole coefficients are then computed by Fourier transformation

$$b_n \; = \; \frac{2}{N F K_n} \sum_{k=0}^{N-1} \Phi(\theta_k) \cos \left(k \cdot n \cdot \frac{2\pi}{N} \right) \qquad (A.2)$$

$$a_n \; = \; \frac{2}{N F K_n} \sum_{k=0}^{N-1} \Phi(\theta_k) \sin \left(k \cdot n \cdot \frac{2\pi}{N} \right) \, . \qquad (A.3)$$

The quantity K_n is a measure of the pick-up coil's sensitivity to a multipole field of order n

$$K_n = \frac{1}{n} \left[\left(\frac{r_2}{r_0} \right)^n - \left(\frac{r_1}{r_0} \right)^n \right] \, . \qquad (A.4)$$

A single coil is insufficient for a precise determination of a small higher-order multipole in the presence of the large main field. As an example consider a dipole magnet with $B_{ref} = 5$ T which contains a $b_{14} = 1 \cdot 10^{-4}$. For pick-up coil radii $r_1 = 0$, $r_2 = 0.8 r_0$ and an angular velocity $\omega = 2\pi \cdot 0.25$ s^{-1} one obtains $U_1 = 50$ V and $U_{14} = 250 \, \mu$V, so a dynamic range of more than 10^5 would be needed. A considerable increase in sensitivity is obtained by using a system of two pick-up coils with internal compensation of the main field. The HERA dipole measuring coil, shown in Fig. A.2a, consists of an outer coil A and an inner coil B, the latter symmetric to the axis, which have equal width and equal number of turns:

$$N_A(r_2 - r_1) = N_B(r_4 - (-r_3)) = N_B(r_4 + r_3) \, .$$

The dipole contribution to the 'compensated' signal $U_{comp} = U_A - U_B$ vanishes. In practice this requires a slight adjustment of the voltages by potentiometers to correct

a) Dipole measuring system

b) Quadrupole measuring system

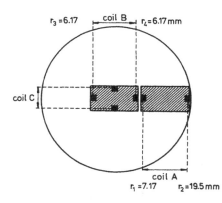
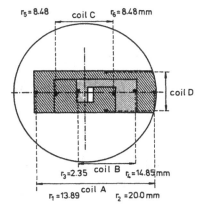

Figure A.2: (a) Dipole pick-up coil system with internal compensation of the dipole component. Subcoil C is used to correct for a possible slight angular tilt of the planes of subcoils A and B. (b) Quadrupole pick-up system with internal compensation of dipole and quadrupole components. Subcoil D is used to correct for slight angular tilts of the planes of subcoils A, B and C.

for small differences in the coil areas. The planes of subcoils A and B may have some angular misalignment. A third coil C, oriented perpendicular to A and B, is used to correct for this possibility.

The dipole field is derived from the voltage U_A alone while for multipoles of order $n > 1$ the compensated signal is used. Equations (A.2), (A.3) are still valid if the sensitivity factor is replaced by

$$K_n = \frac{1}{n}\left[\left(\left(\frac{r_2}{r_0}\right)^n - \left(\frac{r_1}{r_0}\right)^n\right) - \left(\left(\frac{r_4}{r_0}\right)^n - \left(-\frac{r_3}{r_0}\right)^n\right)\right] . \tag{A.5}$$

It is easy to verify that $K_1 = 0$, so the large dipole field is indeed suppressed.

In a quadrupole magnet one wants to suppress the main field B_2 but in addition an apparent dipole component which arises when the pick-up coil is not exactly centred on the magnet axis. A coil system with three subcoils (Fig. A.2b) is capable of performing this task. Coil C is symmetric with respect to the axis of rotation and does not receive an induction from the quadrupole field. The compensated signal is

$$U_{comp} = U_A - U_B - U_C .$$

To cancel the quadrupole field the radii of subcoils A and B have to fulfill the relation

$$(r_2^2 - r_1^2) = (r_4^2 - r_3^2)$$

while cancellation of a possible dipole field component requires

$$(r_2 + r_1) - (r_4 + r_3) - (r_6 + r_5) = 0 .$$

The sensitivity factor is now

$$K_n = \frac{1}{n}\left[\left(\left(\frac{r_2}{r_0}\right)^n - \left(-\frac{r_1}{r_0}\right)^n\right) - \left(\left(\frac{r_4}{r_0}\right)^n - \left(-\frac{r_3}{r_0}\right)^n\right) \right.$$
$$\left. - \left(\left(\frac{r_6}{r_0}\right)^n - \left(-\frac{r_5}{r_0}\right)^n\right)\right] \tag{A.6}$$

and from the above conditions follows $K_1 = K_2 = 0$.

A.2   Systematic errors in harmonic measurements

A.2.1   Feed-down to lower-order multipoles

A displacement of the pick-up coil with respect to the magnet axis leads to apparent multipoles of lower order. In a perfect dipole this is of course irrelevant but in a quadrupole a horizontal displacement generates an artificial normal dipole term b_1, a vertical displacement a skew dipole a_1. This can be seen as follows. Let (x, y) denote the pick-up coil coordinate system and (ξ, η) the magnet coordinate system which is displaced by a vector (u, v). Then

$$\xi = u + x = u + r\cos\theta , \quad \eta = v + y = v + r\sin\theta . \tag{A.7}$$

The field of a pure quadrupole is

$$B_\xi = g\eta , \quad B_\eta = g\xi$$

and the flux through the simple radial pick-up coil of Fig. A.1a is

$$\Phi(\theta) = gN_c l_c \int_{r_1}^{r_2} [(u + r\cos\theta)\cos\theta - (v + r\sin\theta)\sin\theta]dr$$
$$= gN_c l_c[(u\cos\theta - v\sin\theta)(r_2 - r_1) + \frac{1}{2}(r_2^2 - r_1^2)\cos 2\theta] . \tag{A.8}$$

Using again the abbreviation $F = N_c l_c r_0 B_{ref} = gN_c l_c r_0^2$ and the sensitivity factors K_1, K_2 from Eq. (A.4) the flux becomes

$$\Phi(\theta) = F\left[K_1\left(\frac{u}{r_0}\cos\theta - \frac{v}{r_0}\sin\theta\right) + K_2\cos 2\theta\right] . \tag{A.9}$$

The second term is the flux of a pure normal quadrupole field ($b_2 = 1$). It is independent of (u, v) so the displacement has no effect on the main pole:

$$\tilde{b}_2 = b_2 = 1 .$$

The first term describes artificial dipole components. In a simple pick-up coil $K_1 \neq 0$, and then these poles are

$$\tilde{b}_1 = \frac{u}{r_0} b_2, \quad \tilde{a}_1 = -\frac{v}{r_0} b_2 .$$

The quadrupole measuring system described above automatically cancels these unwanted terms. The compensated signal features $K_1 = 0$, see Eq. (A.6), and hence the artificial dipole components vanish as can be seen from Eq. (A.9). In the measurement of a $2n$-pole magnet ($n = 3$ sextupole, $n = 4$ octupole etc) with a simple radial coil we obtain artificial multipoles of order $n - 1$

$$\tilde{b}_{n-1} = (n-1) \cdot \frac{u}{r_0} \cdot b_n, \quad \tilde{a}_{n-1} = -(n-1) \cdot \frac{v}{r_0} \cdot b_n . \qquad (A.10)$$

We illustrate the feed-down effects by a field measurement of the HERA sextupole and quadrupole correction coils (Daum et al. 1989). These coils are wound on top of each other on the beam pipe of the main dipoles. Both are made from single shells and have well-known higher harmonics. The quadrupole layer has non-vanishing 20- and 28-poles $b_{10}/b_2 = 2.5\%$, $b_{14}/b_2 = 0.6\%$, the sextupole coil has a 30-pole $b_{15}/b_3 = 1.5\%$. Their harmonics, determined with a simple pick-up coil, are plotted in Fig. A.3. The displacement of the pick-up coil axis $\Delta r = \sqrt{u^2 + v^2}$ is derived from the observed dipole component in the quadrupole coil ($\Delta r = 0.6$ mm) and then the expected artificial poles of order $n = 9, 13$ in the quadrupole and $n = 2, 14$ in the sextupole are calculated. Fig. A.3 shows that they are in perfect agreement with the measurement.

A.2.2   Generation of higher-order multipoles by irregular axis motion of pick-up coil

The internal compensation of the dipole component has another important implication for the determination of higher order multipoles. Suppose we were given the task to measure a quadrupole with a single pick-up coil. If the pick-up coil axis is not perfectly stable during rotation but performs a small irregular motion about the axis of the magnet then artificial multipoles of orders $n \geq 3$ may be generated. We approximate this motion by a Fourier series in θ.

$$u(\theta) = u_0 + u_{11} \cos \theta + u_{12} \sin \theta + u_{21} \cos 2\theta + u_{22} \sin 2\theta + \ldots \qquad (A.11)$$

and similarly for $v(\theta)$. In the following we confine ourselves to the horizontal axis displacement $u(\theta)$. The term $u_{11} \cos \theta$ modifies the normal quadrupole b_2 slightly while $u_{12} \sin \theta$ generates a skew quadrupole $\tilde{a}_2$. We assume a perfect quadrupole field in our magnet ($b_2 = 1$). Inserting $u(\theta)$ into Eq. (A.9) we find the term

$$u_{12} \sin \theta \cos \theta = \frac{1}{2} u_{12} \sin 2\theta .$$

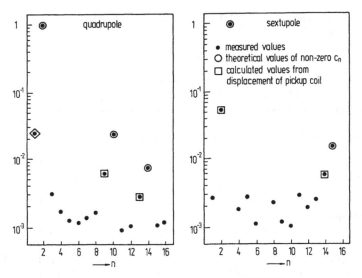

Figure A.3: The measured multipole coefficients $c_n = \sqrt{b_n^2 + a_n^2}$ of a HERA correction coil with an inner sextupole layer and an outer quadrupole layer which are mounted on a common beam pipe. Black dots: measured values, normalized to the quadrupole resp. sextupole field at $r = r_0 = 25$ mm. Open circles: theoretical values for c_{10}, c_{14} in the quadrupole and c_{15} in the sextupole. Open squares: predicted values of c_9, c_{13} in the quadrupole and of c_2, c_{14} in the sextupole. The displacement Δr of the pick-up coil axis was computed from the observed dipole coefficient (diamond) in the quadrupole: $\Delta r = c_1 r_0$.

Using (A.3) and (A.9) we get (for $K_2 \neq 0$)

$$\tilde{a}_2 = \frac{u_{12}}{2r_0} \cdot \frac{K_1}{K_2} \cdot b_2 \; .$$

The term $u_{21} \cos 2\theta$, when inserted into (A.9), introduces a normal sextupole and a normal dipole

$$u_{21} \cos 2\theta \cos \theta = \frac{1}{2} u_{21} (\cos 3\theta + \cos \theta)$$

and so on. The artificial sextupole coefficients of a perfect quadrupole are found to be (Meinke, Schmüser and Zhao 1991)

$$\tilde{b}_3 = \frac{u_{21}}{2r_0} \cdot \frac{K_1}{K_3} \cdot b_2 \, , \quad \tilde{a}_3 = \frac{u_{22}}{2r_0} \cdot \frac{K_1}{K_3} \cdot b_2 \; .$$

A very small amplitude $u_{21} = 10 \,\mu$m leads already to an artificial sextupole in the order of 10^{-4}. Hence the axis motion must be restricted to an amplitude of less 10 μm which is hard to achieve inside a superconducting magnet. Fortunately, the use of a compensating coil system with $K_1 = 0$ eliminates this problem. With the quadrupole pick-up system of Fig. A.2b even high poles like the b_{14} in the HERA quadrupoles can be measured with a precision in the few 10^{-5} range.

The HERA dipoles were measured with the dipole pick-up system of Fig. A.2a. The effects of irregular axis motion can be seen very clearly from Fig. A.4. Here the rms standard deviations of the measured multipole coefficients a_n and b_n of all dipoles made by ABB are plotted against the multipole order. The upper curves (open circles) show the data at low excitation of the magnets (0.23 T) where large persistent-current sextupoles and decapoles ($b_3 = -33 \cdot 10^{-4}$, $b_5 = 12 \cdot 10^{-4}$) are present. The lower curves (full circles) have been obtained at an intermediate field of 2.8 T where the field distortions due to persistent currents and yoke saturation reach a minimum. The rms spread of the multipoles is much lower at 2.8 T than at 0.23 T; the difference is caused by irregular axis motion.

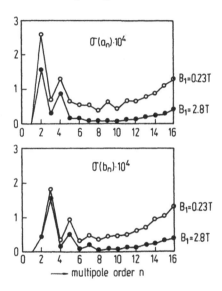

Figure A.4: The rms standard deviations of the multipole coefficients a_n and b_n of all HERA-ABB dipoles at 0.23 T and 2.8 T. The spread of the coefficients a_2, a_4, b_3, b_5 is mainly caused by geometrical imperfections in the dipoles coils. The rms errors of the coefficients with $n > 5$ are significantly larger at 0.23 T than at 2.8 T. The difference is caused by irregular pick-up coil motion in the large persistent-current sextupole and decapole fields present at 0.23 T.

This observation is of relevance for the theoretical determination of the 'dynamic aperture' (maximum stable beam size) of a large hadron storage ring, which is usually accomplished by tracking particles for more than 10^5 turns around the ring. Supercon-

ducting machines are severely affected by the non-linear magnetic field components in
the magnets which are caused by coil imperfections (Chap. 5) and by persistent- and
eddy-current effects (Chaps. 6, 7). The dynamic aperture limitation is strongest at
injection energy because here the beam has its largest extension and the persistent-
current field distortions in the magnets are maximum. If one were to use the multipole
data measured at 0.23 T for the computation of the dynamic aperture of HERA at
injection energy (40 GeV), one would get a far too pessimistic number because the
0.23 T data grossly overestimate the rms spread of higher multipoles. The 'geometric' multipole errors of the magnets must be derived from the intermediate-field
data while the low-field data are used only for determining the persistent-current
multipoles b_3, b_5, b_7.

For further details on multipole measurements we refer to (Walckiers 1992).

References

C. Daum et al., *The superconducting quadrupole and sextupole correction coils for the
    HERA proton ring*, DESY report HERA 89-09 (1989)
R. Meinke, P. Schmüser and Y. Zhao, *Methods of harmonic measurements in the super-
    conducting HERA magnets and analysis of systematic errors*, DESY report HERA
    91-13 (1991)
L. Walckiers, *The harmonic coil method*, Lectures at the CERN school on 'Magnetic
    Measurement and Alignment', Montreux 1992, CERN report 92-05 (1992)

Appendix B

Stretched-wire system for quadrupole measurements

Here we describe the stretched-wire system used for determining the gradient, axis location and field orientation of the HERA quadrupoles. A 100 μm thick copper-beryllium wire is pulled through the magnet parallel to the quadrupole axis (Fig. B.1). Together with an external return wire it forms an induction loop which is connected to a low-noise amplifier and a voltage-to-frequency converter (VFC). The wire can be moved horizontally and vertically with micrometre precision and the resulting flux changes are recorded by scaling the VFC output pulses in an up-down counter. To achieve an accuracy of 10^{-4}, electronic noise at the VFC has to be kept below 100 nV. The connectors and cables are thermally shielded to avoid variations in contact potentials which would normally be in the μV range. Still remaining background voltages lead to a drift of the up-down counter readings which is corrected for as indicated in Fig. B.2. Additional corrections are needed for wire sagging due to gravity and for a slight deflection of the wire in the inhomogenous quadrupole field owing to the diamagnetism of copper-beryllium. These effects are eliminated by measuring at different values of the mechanical tension in the wire and extrapolating the results to infinite tension.

During quadrupole assembly at the manufacturing plants the coils were optically pre-aligned with respect to an external survey target, mounted on the outside of the cryostat vessel. This reference system is chosen as the (x, y) coordinate system of the stretched wire. The quadrupole coordinate system (ξ, η) is in general displaced by a small vector (u, v) and rotated by a small angle α against the (x, y) system, see Fig. B.3a. One has

$$\begin{aligned}
\xi &= (x + u)\cos\alpha + (y + v)\sin\alpha , \\
\eta &= -(x + u)\sin\alpha + (y + v)\cos\alpha .
\end{aligned}$$

In its own system the quadrupole field reads

$$B_\xi = g\eta , \quad B_\eta = g\xi . \tag{B.1}$$

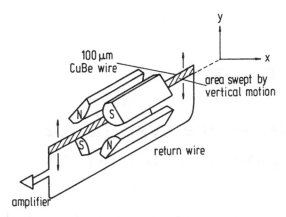

Figure B.1: Schematic view of the stretched-wire system. For clarity a quadrupole with hyperbolic pole shoes is drawn instead of a superconducting coil.

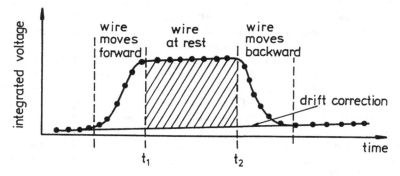

Figure B.2: Example of a magnetic flux measurement with drift correction.

In the stretched-wire system the field components are for $\alpha \ll 1$

$$
\begin{aligned}
B_x &= g\left((y+v)\cos 2\alpha - (x+u)\sin 2\alpha\right) \\
&\approx g\left((y+v) - (x+u)2\alpha\right) , \\
B_y &= g\left((x+u)\cos 2\alpha + (y+v)\sin 2\alpha\right) \\
&\approx g\left((x+u) + (y+v)2\alpha\right) .
\end{aligned}
\tag{B.2}
$$

The movements performed with the wire are sketched in Fig. B.3b. The corresponding flux changes are

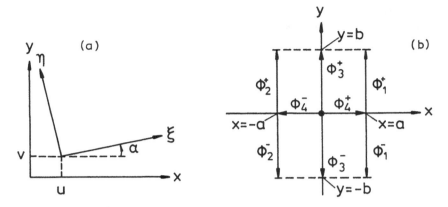

Figure B.3: (a) The coordinate systems of the quadrupole (ξ, η) and the stretched wire (x, y). (b) Movements performed with the wire.

$$
\begin{aligned}
\Phi_1^{\pm} &= l_m \int_0^{\pm b} B_x(a, y) dy \\
&\approx g l_m \left(b^2/2 \pm b[v - (u + a)2\alpha] \right) , \\
\Phi_2^{\pm} &\approx g l_m \left(b^2/2 \pm b[v - (u - a)2\alpha] \right) ,
\end{aligned}
\tag{B.3}
$$

where l_m is the magnetic length of the quadrupole. The symmetric combination

$$
\Phi_1^S = \Phi_1^+ + \Phi_1^- = (g l_m) b^2 = \Phi_2^S
$$

yields the integrated gradient

$$
\int g \, dl \equiv (g l_m) = \Phi_1^S / b^2 .
\tag{B.4}
$$

The field rotation angle α is obtained from the antisymmetric combinations

$$
\begin{aligned}
\Phi_1^A &= \Phi_1^+ - \Phi_1^- = 2 g l_m \, b[v - (u + a)2\alpha] , \\
\Phi_2^A &= \Phi_2^+ - \Phi_2^- = 2 g l_m \, b[v - (u - a)2\alpha] :
\end{aligned}
$$

$$
\alpha = \frac{\Phi_2^A - \Phi_1^A}{8\Phi_1^S} \cdot \frac{b}{a} .
\tag{B.5}
$$

Finally, the displacement of the axis is approximately given by

$$
u \approx \frac{\Phi_4^A}{2\Phi_1^S} \cdot b , \quad v \approx \frac{\Phi_3^A}{2\Phi_1^S} \cdot b .
\tag{B.6}
$$

Further reading

J.DiMarco and J. Krzywinski, *MTF single stretched wire system*, Fermilab report MTF-96-0001 (1996)

Appendix C

Passive superconductor magnetization in beam-pipe coils

In Sect. 9.3.2 we have described the coupled persistent current effects in a dipole which is equipped with beam pipe correction coils. Here we want to show that the 'passive' magnetization effects – the correction coils are not powered but their filaments are magnetized by the external dipole field – can be treated analytically. We consider coils made from single current shells such as depicted in Figs. 4.5 and 4.6. The Tevatron dipole, quadrupole and sextupole correctors and the HERA beam pipe coils are made this way (Fig. 9.8). A dipole coil consists of 2 current sections each covering $\pm 60°$ which are centred at $0°$ and $180°$. A quadrupole comprises 4 sections of $\pm 30°$ which are centred at $0°$, $90°$, $180°$ and $270°$. Finally, a sextupole comprises 6 sections of $\pm 20°$ which are centred at $0°$, $60°$, $120°$, $180°$, $240°$ and $300°$.

The wire arrangement in these coils obeys a top-bottom and a left-right symmetry (remember that no transport current is assumed to flow in the wires). The homogeneous external field, provided by the main dipole, induces the same persistent current-pattern in all filaments. For the computation of the persistent-current fields we can therefore consider four symmetrically located NbTi filaments as shown in Fig. 6.4. There is one important simplification: the magnetizing field is pointing in y direction for all filaments because we have assumed that the correction coils are current-free. Hence the angles α and ϕ in Fig. 6.4 are identical and Eq. (6.6) becomes

$$A^{pair}(r, \theta) = -\frac{2\mu_0 I d}{\pi R} \sum_{n=1,3,\dots} \left(\frac{r}{R}\right)^n \cos(n\theta) \cos\left((n+1)\phi\right) . \qquad (C.1)$$

The top-bottom and left-right symmetry implies that the passively magnetized correction coil wires generate only normal multipole fields of odd order: B_1, B_3, B_5, ... Skew fields and even orders are absent. This applies also for quadrupole or octupole coils although the transport current in such coils generates only multipole fields of even order. The essential difference between transport current and passive superconductor magnetization lies in their different symmetries: in a quadrupole, for example, the transport current alternates in sign when going from one current section to the next (see Fig. 4.6b) while the superconductor magnetization caused by the main dipole field is identical in all four sections.

The expression (C.1) for the vector potential of the four current pairs has to be summed over all filaments in the first quadrant of the xy plane. This is basically equivalent to averaging the term $\cos((n+1)\phi)$ over the angular region covered by the correction coil shells in the first quadrant, and to multiplying the result with the number of filaments per quadrant.

For the dipole layer we obtain

$$
\begin{aligned}
cd(n) &\equiv \langle\cos((n+1)\phi)\rangle = \frac{3}{\pi}\left(\int_0^{\pi/3}\cos((n+1)\phi)\,d\phi\right) \\
&= \frac{3}{\pi(n+1)}\left(\sin\left(\frac{(n+1)\pi}{3}\right)\right) .
\end{aligned}
\tag{C.2}
$$

Similarly, for the quadrupole layer we get

$$
\begin{aligned}
cq(n) &\equiv \langle\cos((n+1)\phi)\rangle = \frac{3}{\pi}\left(\int_0^{\pi/6}\cos((n+1)\phi)\,d\phi + \int_{\pi/3}^{\pi/2}\cos((n+1)\phi)\,d\phi\right) \\
&= \frac{3}{\pi(n+1)}\left[\sin\left(\frac{(n+1)\pi}{2}\right) + \sin\left(\frac{(n+1)\pi}{6}\right) - \sin\left(\frac{(n+1)\pi}{3}\right)\right] .
\end{aligned}
\tag{C.3}
$$

The expression for the sextupole layer is

$$
\begin{aligned}
cs(n) &\equiv \langle\cos((n+1)\phi)\rangle = \frac{3}{\pi}\left(\int_0^{\pi/9}\cos((n+1)\phi)\,d\phi + \int_{2\pi/9}^{4\pi/9}\cos((n+1)\phi)\,d\phi\right) \\
&= \frac{3}{\pi(n+1)}\left[\sin\left(\frac{(n+1)4\pi}{9}\right) + \sin\left(\frac{(n+1)\pi}{9}\right) - \sin\left(\frac{(n+1)2\pi}{9}\right)\right] .
\end{aligned}
\tag{C.4}
$$

The numerical values are listed in the following table

n	1	3	5	7
$cd(n)$	0.413	−0.207	0	0.103
$cq(n)$	0	0.413	0	−0.207
$cs(n)$	0	0	0.413	0

So we arrive at the somewhat surprising result that a single-shell dipole coil layer, magnetized by an external homogeneous field, generates a dipole, a sextupole and a 14-pole field, a quadrupole layer generates a sextupole and a 14-pole, and a sextupole layer generates a decapole. The curves in Fig. 9.9 show that quantitative agreement with the data is obtained.

Passive screening of particle beams from transverse magnetic fields would be desirable in many experimental setups. A NbTi tube with a wall thickness of a few mm might appear the obvious solution but a hard superconductor of this thickness is instable against flux jumping (Chap. 2.4.5). A multilayer shielding made from thin NbTi-Nb-Cu sheets avoids the flux jump instability and has recently be applied to shield transverse fields of up to about 1.5 T (S.L. Wipf, private communication).

Appendix D

Approximate relations for NbTi critical parameters

The critical surface of NbTi has been shown in Fig. 2.11. The critical temperature and upper critical field are defined as the values where the critical surface meets the T resp. B axis.

$$T_c \simeq 9.4 \text{ K} \quad (\text{at } B = 0, J = 0) ,$$
$$B_{c2} \simeq 14.5 \text{ T} \quad (\text{at } T = 0, J = 0) .$$

The upper critical field depends on temperature and can be parametrized as (Lubell 1983)

$$B_{c2}(T) = B_{c2}(0)[1 - (T/9.2)^{1.7}] . \tag{D.1}$$

The critical temperature decreases with increasing field. A good fit is given by the relation (Lubell 1983)

$$T_c(B) = T_c(0)[1 - (B/14.5)]^{0.59} . \tag{D.2}$$

The current-sharing temperature varies between $T_{cs} = T_c(B)$ for zero current and the helium bath temperature T_0 when the transport current approaches the critical current, $I_t \to I_c$. Assuming a linear relationship one obtains

$$T_{cs} = T_0 + [T_c(B) - T_0] [1 - I_t/I_c] . \tag{D.3}$$

As a typical example, we consider a dipole coil, immersed in a helium bath of $T_0 = 4.2$ K, generating a field of $B = 5$ T at 80% of its critical current. Current sharing coupled with heat generation will start for $T \geq T_{cs} = 4.8$ K, so the temperature margin of this coil is only 0.6 K. Such a small temperature increase requires a tiny energy deposition since the heat capacities of all materials except helium are extremely small around 4.2 K. The heat capacity per unit volume of a composite Cu-NbTi conductor can be parametrized as (Lubell 1983)

$$C = \eta \left[(6.8/\eta + 43.8)T^3 + (97.4 + 69.8B)T \right] \quad [\text{J}/(\text{m}^3\text{K})] \tag{D.4}$$

where η is the volumetric proportion of the superconductor in the composite. It takes only about 3 mJ/cm^3 to start producing Joule heating in a NbTi conductor in pool boiling helium.

A useful parametrization of the critical current density in NbTi in the range 4.0 to 4.5 K and 5 to 8 T has been given by Morgan, here quoted after Devred (1992).

$$J_c(B, T) = J_c(5\,\mathrm{T}, 4.2\,\mathrm{K}) \cdot \left(1 - \frac{a_1 \cdot \tilde{T} + a_2 \cdot \tilde{T}^2 - a_3 \cdot \tilde{T}^3}{1 - a_4 \cdot \tilde{B}}\right)$$
$$\cdot \left(\frac{1 - a_5 \cdot \tilde{B}}{1 - a_6 \cdot \tilde{B} - a_7 \cdot \tilde{B}^2}\right) \tag{D.5}$$

with $\tilde{T} = T - 4.2$ K, $\tilde{B} = B - 5$ T, $a_1 = 0.315319$ K^{-1}, $a_2 = 0.01528$ K^{-2}, $a_3 = 0.00161$ K^{-3}, $a_4 = 0.163089$ T^{-1}, $a_5 = 0.231741$ T^{-1}, $a_6 = 0.021249$ T^{-1} and $a_7 = 0.020418$ T^{-2}. In Fig. D.1 this formula is compared with data on SSC cables (Matsumoto et al. 1992). The lower critical field of NbTi is $B_{c1} \approx 0.011$ T at 4.2 K.

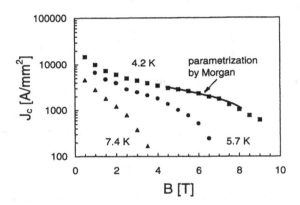

Figure D.1: Critical current densities measured in SSC cables in comparison with formula (D.5).

References

A. Devred, *Quench Origins* in: M. Month and M. Dienes (Eds.), *The Physics of Particle Accelerators*, American Inst. of Physics Conf. Proc. 249, 1992, p. 1262

M.S. Lubell, *Empirical scaling formulas for critical current and critical field for commercial NbTi*, IEEE Trans. **MAG-19** (1983) 754

K. Matsumoto and Y. Tanaka, *Temperature dependence of the critical current density in SSC–type superconducting wires*, Supercollider 4, John Nonte (Ed.), p. 703, Plenum Press 1992

Appendix E

Helium Properties

E.1 Liquid helium

A knowledge of the basic properties of liquid helium is indispensable for the design of magnets made from low-T_c superconductors. Here we mention only a few important properties and refer to Van Sciver (1986) for a detailed description. The phase diagram of ^{4}He (Fig. E.1a) is characterized by four regions: the vapour phase, the two liquid phases He I (normal fluid) and He II (super fluid) and finally the solid phase. The vapour-pressure curve ends at the critical point $T_{crit} = 5.2$ K, $p_{crit} = 2.26$ bar. At 4.2 K the vapour pressure is 1 bar $= 10^5$ Pa. Lower temperatures are reached by pumping on the liquid. At $T = 2.172$ K, a transition to the superfluid phase takes place. This temperature is called the 'λ point' because the specific heat of liquid helium exhibits an anomaly resembling the greek letter λ, see Fig. E.1b. The specific heat of copper is shown for comparison. Like for all solids it vanishes rapidly in the limit $T \rightarrow 0$. At 4.2 K the heat capacity per unit volume of liquid helium is three orders of magnitude larger than that of copper. Therefore, filling the voids in the Rutherford cable with liquid helium increases the overall heat capacity of the cable by nearly two orders of magnitude.

Many of the fascinating phenomena observed in helium II can be understood in terms of the so-called *two-fluid model*. Below the λ point the liquid consists of two interpenetrating phases, a superfluid phase of density ρ_s and a normal fluid of density ρ_n. The total density $\rho = \rho_s + \rho_n$ is almost independent of temperature. The superfluid proportion increases rapidly with decreasing temperature. The superfluid phase features zero viscosity, resulting in no resistance when flowing in tubes (up to a critical velocity), zero entropy and zero heat capacity. Therefore it cannot transport heat. Heat is carried by the normal component, flowing from the warm to the cold side. This is accompanied by a counterflow of the superfluid phase from the cold to the warm side where this phase is partly transformed to normal fluidity. A detailed account on superfluidity is found in (Tilley and Tilley 1990).

Important properties of helium liquid and vapour can be calculated with the

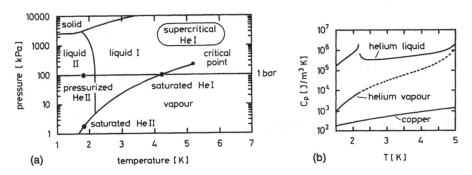

Figure E.1: (a) Phase diagram of ^{4}He. (b) Specific heat of helium liquid and the anomaly at $T_\lambda = 2.17$ K. The heat capacities of helium vapour and of copper are shown for comparison. In this low-temperature range the specific heats of solids becomes extremely small, so helium is the only substance with an appreciable heat capacity.

computer program HEPAK[1]. In Fig. E.2 the computed density and viscosity of liquid helium in equilibrium with vapour are plotted as a function of temperature.

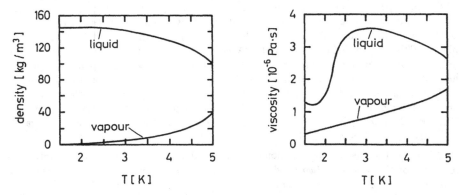

Figure E.2: Computed density and viscosity of liquid helium in equilibrium with vapour as a function of temperature.

The heat conductivity of liquid He I amounts to 0.018 W/(m K) in the range from 3.6 to 5.0 K. The heat of vaporization is around 23 J/g between 1.5 and 4 K and drops to zero when the critical point at 5.2 K is approached. For cooling with He I of 4–4.5 K different methods are in use:

---

[1]HEPAK version 3.01, V. Arp, R.D. McCarty, B.A. Hands, copyright by CRYODATA.

(a) The superconductor may be completely immersed in a bath with the liquid in thermal equilibrium with vapour (two-phase cooling). Heat generated in the cable is transported through the coil into the helium bath. The heat is removed by evaporating liquid and by convection. At the interface between metal and liquid nucleate boiling (bubble formation) or film boiling (vapour film at the boundary) may occur, depending on the amount of heat produced. For heat flows up to 6000 W/m² a temperature difference of less than 0.6 K can be maintained between the metal surface and the helium bath, see Fig. E.3a. When the heat flux is increased beyond this

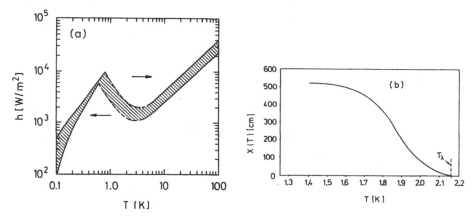

Figure E.3: (a) Heat transfer from a metal surface to normal liquid helium, boiling at 4.2 K and 1 bar. The shaded area indicates the spread of the data (after Wilson 1983, by permission of Oxford University Press). (b) The function $X(T)$ needed to compute the heat flux in superfluid helium at 1 bar (Bon Mardion et al. 1979). (© 1979 Butterworth, Heinemann)

value there is a spontaneous transition from nucleate boiling to film boiling resulting in a drastic rise of the surface temperature to values of 20 K or more which makes film boiling practically useless for magnet cooling. The transition is characterized by a large hysteresis. To recover nucleate boiling the heat flux must be reduced to about 1000 W/m².

(b) The superconductor may be cooled with liquid helium at a pressure above the critical pressure of 2.26 bar. Then liquid and vapour can no longer be distinguished. One speaks of single-phase cooling with supercritical helium. The heat is removed from the superconductor by heat conduction but there is no bubble formation in the liquid, not even at the hottest spot of the coil. This is a very safe way of cooling a superconducting coil provided the single-phase helium is kept at constant temperature via heat exchange with a two-phase mixture of liquid and vapour.

(c) In many magnets the conductor is cooled by forced flow of helium through tubes which are in close thermal contact with the cable. The tubes may be an inte-

gral part of the coil, if for instance the superconductor is wrapped around the tube
or if the cable is surrounded by the tube (cable-in-conduit conductor). This cooling
scheme is mainly used for large detector and fusion magnets. Because of the pressures
needed to maintain the forced flow the helium is usually in the supercritical state.

For temperatures in the 2 K range or lower, cooling with superfluid helium is utilized. Again there are different possibilities. Two-phase bath cooling with saturated
vapour in equilibrium with the liquid is frequently applied for superconducting RF
cavities. The disadvantage is that the cryogenic system operates at subatmospheric
pressure and that leaks to air may cause a contamination. For the LHC magnets,
single-phase cooling with a pressure of about 1 bar is foreseen. Here heat exchange
with a two-phase superfluid helium system is needed.

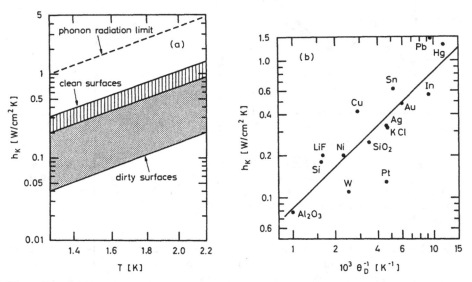

Figure E.4: (a) Experimental values for the Kapitza conductance between copper and He II
for clean and dirty surfaces (Snyder 1969, 1970). (b) The Kapitza conductance for various
solids, plotted against the inverse Debye temperature. The largest observed values are
shown (Challis 1968). (a: © 1970 Butterworth, Heinemann)

The heat conductivity of helium II is very high. For a column of length L whose
left end is at the lower temperature T_1 and the right end at the higher temperature
T_2 (both must be below T_λ), the heat flux $\dot{q}$ is given by the formula (Bon Mardion et
al. 1979)

$$\dot{q} = \dot{q}_0 \left[(X(T_1) - X(T_2))/L \right]^{0.294} \tag{E.1}$$

with $\dot{q}_0 = 1$ W/cm². The length L is given in cm. The function $X(T)$, determined from measurements at 1 bar, is plotted in Fig. E.3b.

It is instructive to compare the heat transport in superfluid helium to that in very pure copper. Let us assume $T_1 = 1.8$ K, $T_2 = 2$ K and $L = 100$ cm. Then, using Fig. E.3b, the heat flux through the helium column is $\dot{q}=1.37$ W/cm^2. For copper of 99.98% purity the heat conductivity in the 2 K range is $\lambda \approx 120$ W/m K. The computed heat flux through a 100 cm long Cu column exposed to a temperature difference of 0.2 K is 0.0024 W/cm^2. For very pure copper (99.999% purity) the heat flux rises to 0.0135 W/cm^2. These numbers demonstrate that helium II is a far better heat conductor than any metal.

Heat flux from a metal surface into superfluid helium requires a temperature gradient according to the relation

$$\dot{q} = h_K \Delta T .$$ (E.2)

The quantity h_K is called *Kapitza conductance*. For copper it is plotted in Fig. E.4a as a function of temperature. The Kapitza conductance depends on the type of metal and the cleanliness of the surface, clean surfaces having a higher h_K and hence a lower temperature drop. The theoretical interpretation is based on refraction and reflection of phonons at the metal-helium interface. That phonon-wavelength matching plays a role can be seen from Fig. E.4b which shows that the Kapitza conductance scales with the inverse Debye temperature of the solid.

E.2   Breakdown voltage in helium gas

High voltages in the kV range may occur inside superconducting coils or against ground potential in the case of quenches or rapid current switch-off. The cable and ground insulation is made from a material with high dielectric strength, for instance Kapton, but it is usually not hermetically closed. At certain places the path length for electric discharge may be only a few millimetres or centimetres. This is sufficient in liquid helium whose breakdown voltage is comfortably high due to its high density. The situation is quite different when helium evaporates and heats up during a quench. For gaseous helium at room temperature the breakdown voltage is at least a factor of five lower than in air. This is a particular worry for potential wires which are guided through a tube from the cold coil to electrical feedthroughs at the outer vacuum vessel. In several HERA quadrupoles electric discharges were observed at the room-temperature feedthroughs. The breakdown voltage in helium gas is plotted in Fig. E.5a as a function of pressure and in Fig. E5.b as a function of gap width. Figure E.6 shows the breakdown voltage in a helium-nitrogen mixture as a function of composition. The data are well parametrized by the function

$$V = \sqrt{f_{He}}\, V_{He} + \sqrt{f_{N_2}}\, V_{N_2}$$

with $V_{He} = 1$ kV, $V_{N_2} = 10$ kV and f_{He}, f_{N_2} being the volumetric proportions of He and N_2.

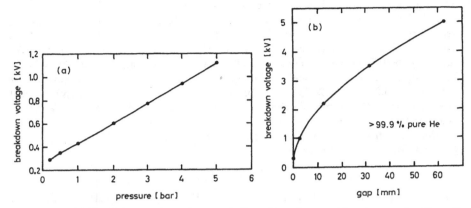

Figure E.5: (a) Breakdown voltage in pure helium for a 1 mm wide gap as a function of helium pressure. The measurements were made by R. Lange of DESY. (b) Breakdown voltage versus gap width in 99.9% pure helium of 1.05 bar at room temperature (unpublished Fermilab data).

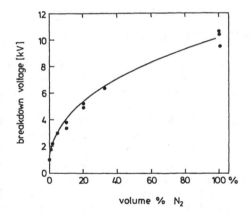

Figure E.6: Breakdown voltage in a helium-nitrogen mixture as a function of the volume percentage of N_2 for a gap width of 4.8 mm and a pressure of 1.05 bar (Fermilab data).

It is interesting to note that a few % admixture of nitrogen or air is sufficient to double the breakdown voltage. If the voltages in helium can be kept below 1 kV during all phases of operation, including a worst-case quench in the magnet chain, a test voltage of 5 kV in dry air at room temperature is the absolute minimum. In tests with gaseous helium one should keep in mind that a small contamination with air raises the breakdown voltage so that the insulation may appear better than it actually is.

References

G. Bon Mardion, G. Claudet, P. Seyfert, *Practical data on steady state heat transport in superfluid helium at atmospheric pressure*, Cryogenics **19** (1979) 45

L. J. Challis, *Experimental evidence for a dependence of the Kapitza conductance on the Debye temperature of a solid*, Phys. Lett. **26A** (1968) 105

N. S. Snyder, *Thermal conductance at the interface of a solid and helium II (Kapitza conductance)*, National Bureau of Standards Technical Note 385, 1969; *Heat transport through helium II: Kapitza conductance*, Cryogenics **10** (1970) 89

D.R. Tilley and J. Tilley, *Superfluidity and Superconductivity*, Third Edition, Institute of Physics Publishing Ltd, Bristol 1990

S. W. Van Sciver, *Helium Cryogenics*, Plenum Press, New York 1986

M. N. Wilson, *Superconducting Magnets*, Clarendon Press, Oxford 1983

Appendix F

Material properties of Kapton

Kapton has turned out the best choice for cable and coil insulation of accelerator magnets with NbTi superconductor. Generally it is used in two forms: as a 25 μm thick and about 10 mm wide film in combination with epoxy-impregnated glass-fibre tape for cable insulation and as a 125 μm thick film of larger width for coil ground insulation. Recently Kapton CI cable insulation system with polyimide adhesive coating has been applied, for instance in the RHIC magnets. Some important properties of Kapton film are listed in Table F.1. The data are taken from (Du Pont 1980). The stress-strain curves exhibit a strong temperature dependence (Fig. F.1a). Tensile and yield strength increase considerably at low temperature (Fig. F.1b) while the modulus of elasticity exhibits a moderate increase from 2.4 GPa at 300 K to about 3.8 GPa at 4 K. Like many plastics, Kapton film suffers from creep under heavy load, see Fig. F.2. The creep rate is reduced at low temperature. In an experiment at CERN a 36-mm-thick stack of 24 cable sections with Kapton and glass-tape insulation, such as used in the HERA magnets, was polymerized with a pressure of 90 MPa and then subjected to a long-term creep test under a load of 100 MPa at room temperature. A logarithmic creep was observed with a reduction in stack thickness of 30 μm within the first day and of 60 μm in the next 100 days (H. Kummer, private communication). Despite its excellent electrical and satisfactory mechanical properties it must be noted that Kapton foil is vulnerable to damage by sharp edges and that it may contain pin holes. It is therefore recommendable to use a multilayer insulation to prevent electrical shorts at high voltage. Excellent electrical and mechanical strength is obtained in glass-Kapton or glass-Kapton-glass sandwich foils which are frequently used for cable insulation at magnet-interconnections. The mechanical strength of Kapton vanishes for temperatures above 1088 K. More details on the temperature dependence of Kapton film properties are found in (Hatfield 1970).

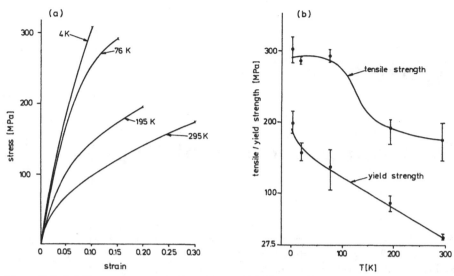

Figure F.1: (a) Stress versus strain for a 25 μm thick Kapton film at different temperatures. (b) Tensile and yield strength of Kapton film as a function of temperature. (After Hatfield (1970)).

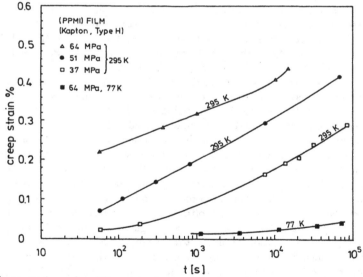

Figure F.2: Creep strain under different loads as a function of time at 295 K and 77 K (Hatfield 1970).

Table F.1: Properties of 25 μm thick Kapton film.

		78 K	296 K	473 K
mechanical properties				
	ultimate tensile strength [MPa]	241	172	117
	yield point at 3 % [MPa]		69	41
	ultimate elongation [%]	2	70	90
	tensile modulus [GPa]	3.5	3.0	1.86
	density [g/cm^3]		1.42	
thermal properties				
	coefficient of linear expansion [K^{-1}]		$2.0 \cdot 10^{-5}$	
	thermal conductivity [W/mK]		0.155	0.178
	specific heat [J/gK]		1.09	
electrical properties				
	dielectric strength [V/μm] at 60 Hz		276	
	dielectric constant at 1 kHz		3.5	
	dissipation factor at 1 kHz		0.0025	
	volume resistivity [Ωm] at 125 V		$1 \cdot 10^{16}$	
chemical properties				
	radiation resistance [Gy]		$4 \cdot 10^7$	

References

Du Pont manual 1980, *Kapton polyimide film, summary of properties*

R. F. Hatfield, Lawrence Livermore Laboratory, *Interim report 275.05-70-1 on some thermo-physical properties of Dacron thread, Mylar and Kapton film from 1.8 to 300 K*, Cryogenic Division, Institute for Basic Standards, NBS Boulder, Colorado 1970

Appendix G

Collar and yoke material properties

G.1 Stainless steel

Stainless steel is the most common material for collaring the coil. During cooldown it shrinks somewhat less than the coil, so a higher pre-compression is needed at room temperature to achieve the desired coil pre-stress in the superconducting state. Great care must be taken that the stainless steel type chosen does not become magnetic either by welding, by cold work (e.g. stamping) or upon cool-down. The magnetic properties of stainless steels depend on the chemical composition. For a number of important stainless steels it is listed in Table G.1.

The first eight entries in the table are steels according to the German DIN standard followed by eight steels according to the American AISI/SAE standard. There is a rough equivalence between the standards, for instance 1.4301≃ 304, 1.4306 ≃ 304L, 1.4311≃ 304N, 1.4401≃ 316, 1.4404 (1.4435) ≃ 316L, 1.4406 ≃ 316N, 1.4429≃316LN.

Whether a stainless steel may become magnetic is determined by the amount of δ ferrite formation. The δ ferrites are normally converted to austenite by annealing but they may re-appear after cold work, welding or during cooldown. To determine the δ ferrite content, one uses the 'Schaeffler' diagram (Fig. G.1), in which the nickel-equivalent, promoting the austenite, is plotted versus the chromium-equivalent, promoting the ferrite (Larbalestier 1977). The nickel and chromium equivalents are defined as follows:

$$
\begin{aligned}
\text{Ni-equivalent} &= f_{Ni} + 0.11 f_{Mn} - 0.0086 f_{Mn}^2 + 18.4 f_N + 24.5 f_C \\
\text{Cr-equivalent} &= f_{Cr} + 1.21 f_{Mo} + 0.48 f_{Si} \,,
\end{aligned}
$$

where f_{Ni} denotes the nickel proportion in the alloy etc.

Stainless steels above the 0% δ-ferrite line remain non-magnetic at 4 K while the steels below the 5% δ ferrite line (towards a higher proportion of δ ferrites) are questionable for application in the collars of a superconducting coil or in the beam pipe. Due to the appreciable range in chemical composition (see Table G.1) each steel type covers a wide range in the Schaeffler diagram, extending from the 'good' region above the 0% δ-ferrite line to the 'bad' region below the 5% line. Hence it is not sufficient to choose a steel type by its name only. In addition the permitted range in chemical composition has to be specified and the working point in the Schaeffler diagram has to be determined from a chemical analysis of the material. Materials containing nitrogen are particularly suitable for use in the magnetically sensitive region close to the particle beam. The German type DIN 1.4429 and the American type 316LN belong into this class.

Table G.1: Chemical composition of stainless steels in weight %. The numbers quoted for the elements C, Si, Mn, P and S are upper limits.

type	C	Si	Mn	P	S	Cr	Mo	Ni	N
1.4301	0.07	1.0	2.0	0.045	0.03	17.0–19.0	—	8.5– 11.0	—
1.4306	0.03	1.0	2.0	0.045	0.03	18.0–20.0	—	10.0– 12.5	—
1.4311	0.03	1.0	2.0	0.045	0.03	17.0–19.0	—	8.5– 11.5	0.12–0.22
1.4401	0.07	1.0	2.0	0.045	0.03	16.5–18.5	2.0–2.5	10.5–13.5	—
1.4404	0.03	1.0	2.0	0.045	0.03	16.5–18.5	2.0–2.5	11.0–14.0	—
1.4406	0.03	1.0	2.0	0.045	0.03	16.5–18.5	2.0–2.5	10.5–13.5	0.12–0.22
1.4429	0.03	1.0	2.0	0.045	0.03	16.5–18.5	2.5–3.0	12.0–14.5	0.14–0.22
1.4435	0.03	1.0	2.0	0.045	0.03	16.5–18.5	2.5–3.0	12.5–15.0	—
304	0.08	1.0	2.0	0.045	0.03	18.0–20.0	—	8.0–10.5	—
304L	0.03	1.0	2.0	0.045	0.03	18.0–20.0	—	8.0–12.0	—
304N	0.08	1.0	2.0	0.045	0.03	18.0–20.0	—	8.0– 10.5	0.10–0.16
304LN	0.03	1.0	2.0	0.045	0.03	18.0–20.0	—	8.0–10.5	0.10–0.15
316	0.08	1.0	2.0	0.045	0.03	16.0–18.0	2.0–3.0	10.0–14.0	—
316L	0.03	1.0	2.0	0.045	0.03	16.0–18.0	2.0–3.0	10.0–14.0	—
316N	0.08	1.0	2.0	0.045	0.03	16.0–18.0	2.0–3.0	10.0–14.0	0.10–0.16
316LN	0.03	1.0	2.0	0.045	0.03	16.0–18.0	2.0– 3.0	10.0–14.0	0.10–0.30
Nitronic 40	0.08	1.0	8.0-10.0	0.06	0.03	19.0–21.5	—	5.5– 7.5	0.15–0.40

Fe-Cr-Ni steels may transform from austenite to martensite below a transition temperature T_M. This is accompanied with the appearance of ferromagnetic regions.

$$T_M[K] = 1578 - 61.1 f_{Ni} - 41.7 f_{Cr} - 33.3 f_{Mn} - 27.8 f_{Si} - 1667 f_{(C+N)} - 36.1 f_{Mo} \ .$$

The uncertainty in T_M is $\pm$ 50 K. Plastic deformation may cause the transition to martensite to take place already at temperatures 300 – 400 K higher than T_M.

A partly magnetic material may seriously affect the field quality of a quadrupole as shown in Fig. G.2. As a typical example for the mechanical properties of stainless steels at low temperature, the ultimate tensile strength and the yield strength (equivalent to the German σ_{02} value) of steel type 316 are plotted in Fig. G.3. The modulus of elasticity is about 200 GPa. During cooldown from 300 K to 4 K the yield strength of the material increases from 275 to about 670 MPa while the thermal conductivity decreases from 15 to 0.2 W/(m K). The thermal contraction amounts to 0.265% (see Fig. G4). The electrical resistivity changes only little, from $7.5 \cdot 10^{-7}$ to $5.3 \cdot 10^{-7} \Omega m$.

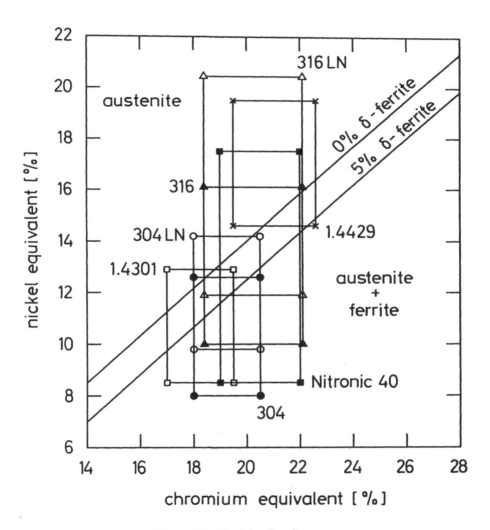

Figure G.1: The Schaeffler diagram.

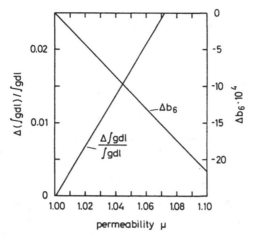

Figure G.2: Computed influence of the magnetic permeability of the stainless steel collars on the integrated gradient and the 12-pole of a HERA quadrupole (J. Perot, private communication).

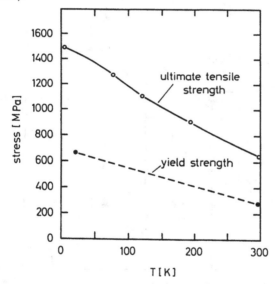

Figure G.3: Temperature dependence of ultimate tensile strength and yield strength of stainless-steel 316 (Handbook 1974).

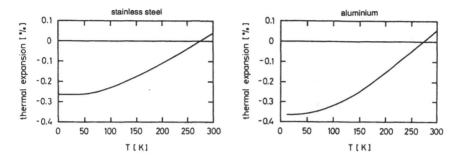

Figure G.4: Thermal expansion of stainless steel and aluminium between liquid helium and room temperature.

G.2   Aluminium alloy

Aluminium alloy is fully non-magnetic and has the additional advantage that it shrinks more upon cooldown than stainless steel (Fig. G.4), namely practically by the same amount as the coil. Therefore, an aluminium-collared coil requires no excess in pre-compression at room temperature to guarantee the specified pre-stress at 4 K. Furthermore aluminium alloy is generally cheaper than stainless steel. For these reasons the alloy AlMg4.5Mn, G35 (German Werkstoff-Nummer 3.3547) was chosen for the collars in the HERA dipoles. The chemical composition is shown in Table G.2. An equivalent alloy is the American standard 5083-0.

Table G.2: Chemical composition in weight % of AlMg4.5Mn, G35

Si	Fe	Cu	Mn	Mg	Cr	Zn	Ti
≤ 0.4	≤ 0.4	≤ 0.1	0.4–1.0	4.0–4.9	0.05–0.25	≤ 0.25	≤ 0.15

At room temperature the ultimate tensile strength is 345–405 MPa and the yield strength 270 MPa. In Fig. G.5 these quantities are shown as a function of temperature (Handbook 1974). In the R&D phase of the HERA dipoles there was some concern that aluminium-alloy might suffer from fatigue under repeated load cycles. A 1-m-long dipole coil was subjected to 500 cycles from minimum to maximum field without any indication of weakening in the collar material. Moreover, since the commissioning of HERA the 416 dipole magnets have been cycled many hundred times without measurable degradation.

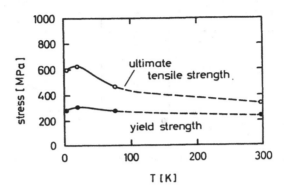

Figure G.5: Temperature dependence of ultimate tensile and yield strength of aluminium-alloy 5083.

G.3   Yoke iron properties

The material used for iron yokes is a low-carbon mild steel whose magnetic properties like coercive force and permeability must be kept within tight limits, especially at saturation.The chemical composition is decisive and has to be controlled during production. As an example we present in Table G.3 the specificied chemical composition of the iron for the RHIC magnets. The yield strength (0.2% deformation) is $\geq$ 224

Table G.3: Chemical composition of RHIC yoke iron (A.F. Greene, private communication).

C	Si	Al	N	Mn	sum of others
$\leq$0.006%	$\leq$0.050%	$\leq$0.080%	$\leq$0.005%	0.100–0.300%	$\leq$0.2%

MPa at room temperature. The coercive force measured after an excitation to 7960 A/m (100 Oerstedt) is $H_c \leq$ 140$\pm$ 20 A/m. The relative permeability at 79.6 A/m is $\mu \geq$ 500. For data on the HERA yoke iron see Fig. 10.5.

References

Handbook on materials for superconducting machinery, MCIC-HB-04, Metals and Ceramics Information Center, Batelle Columbus Laboratories, Nov. 1974

D.C. Larbalestier, *Selection of stainless steel for the Fermilab energy doubler/saver magnets*, Fermilab note TM-745 1630.000, Oct. 1977

Index